Contents

Sheep Management and Wool Production

David Crean
Geoff Bastian

INKATA PRESS

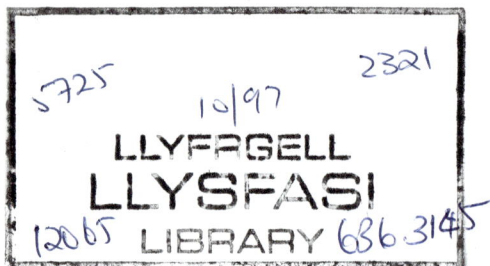

INKATA PRESS

A division of Butterworth-Heinemann Australia

Australia
Butterworth-Heinemann, 18 Salmon Street, Port Melbourne, 3207

Singapore
Butterworth-Heinemann Asia

United Kingdom
Butterworth-Heinemann Ltd, Oxford

USA
Butterworth-Heinemann, Newton

National Library of Australia Cataloguing-in-Publication entry

Crean, David.
Sheep management and wool production.

Includes index.
ISBN 0 7506 8915 3.

1. Wool - Australia. 2. Sheep - Australia. 3. Sheep
ranches - Australia - Management. I. Bastian, Geoff, II.
Title. (Series : Practical farming).

636.3145

Typeset by Ian MacArthur, Hornsby Heights, NSW.

Printed in Singapore by Chung Printing

CHAPTER 1

Introduction

The Australian wool industry

The world sheep population is currently about 1100 million. Most sheep are found in Australia, China, the countries of the former Soviet Union, and New Zealand. In countries other than Australia and New Zealand sheep are used for local meat and wool production, with little being exported.

Table 1.1 *World sheep and wool production (ABARE Australian Commodity Statistics 1995)*

Country	Number (millions)	Wool (thousands of tonnes)	Meat (thousands of tonnes)
Australia	125	726	640
New Zealand	51	272	500
China	110	260	715
Former USSR	89	298	620
World total	1100	2636	6960

Due to our relatively small consumption and great numbers of sheep, Australia and New Zealand have a large effect on the supply of wool and sheep meat products on the world market. Australia and New Zealand dominate the world lamb and mutton trade, with 300 000 tonnes and 400 000 tonnes exported respectively. Most of this product is imported by the European Union, USA, Japan and the Middle East. Australia also exports a significant number of live sheep (5.6 million in 1994) to the Middle East.

Over half of the world's supply of Merino wool and greater than 70 per cent of world apparel wool exports is produced in Australia. Apparel wool is wool made to produce garments. Most Merino wool produced in Australia is between 19 and 26 microns and used for this purpose. Many other wool

7

producing countries produce crossbred wool over 26 microns or carpet wool from 33 to 40 microns.

Australia's main wool importing customers are China, Italy, Japan and France. The pattern of Australian wool exports has changed in the last few years: sales to the former Soviet Union have ceased, there has been a reduction in exports and reliance on Japan and wool sales into Korea and China have expanded.

Table 1.2 *Destination of Australian wool exports (thousands of tonnes: ABARE Australian Commodity Statistics 1995)*

Country	1987-88	1991-92	1994-95
Italy	88	121	109
France	83	100	70
China	97	114	176
Japan	178	149	88
Former USSR	90	32	0
Total	925	928	812

Current situation

Australia presently has approximately 125 million or 10 per cent of world sheep numbers. Within Australia New South Wales has the largest number of sheep with 43.75 million (35 per cent), followed by Western Australia with 31.25 million (25 per cent), Victoria with 23.75 million (19 per cent), South Australia with 12.5 million (10 per cent) and Queensland with 10 million (8 per cent).

A large portion of the national flock is run in mixed farming areas where there are alternative enterprises to sheep production possible. Actual sheep numbers in any one year will vary significantly. This depends on the seasonal conditions in each area, the price of wool and other sheep products and the price and cost structure of other rural enterprises competing with sheep for available area.

For example, sheep numbers were a low 120 million in 1981–82 due the severity of drought conditions in south-eastern Australia. These numbers reached a peak of 175 million in 1990, with a response to high wool prices as sheep replaced other enterprises. Numbers have now fallen to an estimated 125 million in 1996 because of low wool prices and an expansion of other enterprises.

Ewes make up approximately half of the Australian flock, and each year produce about 70 lambs per 100 ewes joined. These lambs are used for replacements in the existing flock, about half are sold for slaughter and the remainder are sold to other producers to run as woolgrowers.

The average weight of wool cut per head in Australia is 4.5 kilograms. This figure varies with the breed and age of sheep.

Sheep and wool products make a significant contribution to the Australian economy. In 1994–95, lamb and mutton exports were worth $500 million, the live sheep trade was valued at $150 million and wool exports at $4000 million. This represents export earnings from the Australian sheep industry of $4.65 billion.

Wool producers

There are three broad types of sheep used in Australia: wool breeds, meat breeds and dual purpose. The many different breeds and strains of sheep used in Australia can be placed into one of these groups. The major breed of sheep is the Merino making up approximately 75 per cent of the national flock.

Australian Merino

The Merino breed consists of four strains that are distinct in wool type and body size: fine wool, medium wool, South Australian and poll Merinos.

Fine wool

The fine wool or Saxon strain is a small framed animal (mature ewe body weight is about 45 kilograms) with fine and superfine wool in the 17 to 19 micron range. The wool is very bright and highly resistant to fleece rot caused by high rainfall. The sheep is well covered around the face and legs and has an annual wool cut of 2.5 to 4 kilograms. Most fine wool sheep are run in the high rainfall areas of New South Wales, Victoria and Tasmania. Fine wool Merinos make up about 5.0 per cent of the Australian Merino flock.

Figure 1.1 *Fine wool Merino*

Figure 1.2 *Medium wool Merino*

Medium wool

There are two strains of medium wool Merinos used in Australia. The medium non-Peppin (or Spanish) strain is a medium framed sheep (mature ewe body weight of 50 kilograms) with medium wool in the 20 to 22 micron range. The wool is bright, resistant to fleece rot and has an annual wool cut of 5 to 7 kilograms. Most Spanish strain Merinos are run in the high rainfall area and sheep/wheat zone of New South Wales.

The Peppin strain is a medium framed sheep with wool in the 21 to 23 micron range. Annual wool cut is 5 to 7 kilograms. Peppin strain sheep are run across large areas of New South Wales, Victoria, Queensland and Western Australia and comprise approximately half of Merinos run in Australia.

Figure 1.3 *South Australian Merino*

Figure 1.4 *Poll Merino*

South Australian

The South Australian Merino is a large framed sheep (mature ewe body weight of 55 kilograms) with wool in the 23 to 26 micron range. Annual wool cut is 5 to 7 kilograms. South Australian Merinos are more open faced than other strains, have less wool down the legs and less body wrinkle. This strain of Merino is the predominant type in the pastoral areas of New South Wales, South Australia and Western Australia, and makes up 40 per cent of the Australian Merino flock.

Poll Merinos

The poll Merino ram (a ram without horns) was developed in Australia. Poll rams are easier to handle, are less likely to become tangled in fences and are less likely to get fly strike in the poll area. Several studs produce only poll rams and many studs have a poll "family" in an otherwise horned stud.

Carpet wool

Carpet wool is a specialty product that requires a large percentage of medullated (or hollow) fibres to give the carpet the ability to retain its shape under constant wear. Carpet wool sheep in Australia are based on the Romney Marsh breed and include the Drysdale and Tukidale. Carpet wool sheep produce a fleece that is chalky white, without crimp and with a high percentage of medullated fibres. The fleece grows about 30 centimetres per year and the ewe produces 5 to 7 kilograms of 35 to 45 micron wool annually in two shearings. Carpet wool breeds also produce an acceptable prime

Figure 1.5 *Carpet wool sheep (Tukidale)*

lamb carcase. Carpet wool sheep are mainly confined to the higher rainfall areas of Australia.

Dual purpose breeds

Dual purpose breeds have been developed in Australia in an attempt to improve the carcase qualities of the Merino, while retaining its wool characteristics. The two main dual purpose breeds used in Australia are the Corriedale and the Polwarth.

The Corriedale

Corriedales were developed by crossing Merino ewes with a Lincoln ram, a large, coarse wool breed, and then interbreeding the progeny. Corriedales

Figure 1.6 *Polwarth*

produce a good style fleece of 5 to 7 kilograms per year in the 25 to 30 micron range. The ewes are large framed (mature weight of 60 kilograms) and produce either a prime lamb as a pure bred, or are crossed with a meat breed to improve carcase characteristics. Most Corriedales are run in the higher rainfall areas.

Polwarth

The Polwarth is a Corriedale crossed back to a Merino ram.

Compared to a Corriedale, wool quality is improved but carcase quality is reduced and ewe fertility is lower. Carcase quality traits are superior to Merino lambs. Polwarth ewes are medium framed (mature ewe body weight of 50 kilograms) and produce 5 kilograms of 22 to 24 micron wool per year. Most Polwarths are run in the colder, wetter areas of Victoria and Tasmania.

Meat breeds

First Cross ewe

The prime lamb industry in Australia is based on the Merino-Border Leicester cross ewe. This ewe is commonly referred to as a "crossbred" ewe. The ewe is large framed (mature ewe body weight of 55 to 60 kilograms) and pro-

Figure 1.7 *First Cross ewe*

duces 4 kilograms of 26 to 30 micron wool per year. This cross is especially suitable for the production of prime lambs because it exhibits:

- High fertility levels (marking percentages are usually around 120).
- Large frame with an ability to lamb from larger meat bred sires.
- Excellent mothering ability, actively protecting the lamb.
- High milk production meaning faster lamb growth rates and allow finished lambs to be sold earlier.

Terminal sires: British longwool

The main sire from this group used in Australia is the Border Leicester, used to produce First Cross ewes. This breed is a large framed animal (mature ewe body weight of 65 kilograms) with wool around 36 microns. (A terminal sire is the sire used to mate with the First Cross ewe to produce a prime lamb.)

Terminal sires: shortwools

The main shortwool or Downs breeds used as terminal sires are the Poll Dorset, Suffolk and South Suffolk and the Southdown. Each of these breeds produces lambs that vary in the age of maturity (the weight at which the animal begins to lay down fat) and so are suited to producing carcases of different weights.

Table 1.3 *Comparison of shortwool breeds*

Breed	Carcase weight	Comments
Southdown	12 to 16 kg	Early maturing. Light lambs.
Poll Dorset	16 to 19 kg	Medium to heavy lambs. Rams from some sources are better suited to producing heavier carcase weights up to 22 kg.
Suffolk	18 to 20 kg	Late maturing breed. Also suitable for heavy carcase weights up to 23 kg.

Other terminal sires

In the past five years, other breeds have been imported into Australia in an attempt to satisfy a growing domestic and export market for heavy, lean lamb carcases. The most notable of these is the Texel. This breed provides a fast growing lamb that produces a heavy carcase over 22 kilograms with low fat cover.

Production systems

There is a wide range of sheep production systems used across Australia in environments from cold tablelands country receiving over 1000 millimetres of rainfall per year to hot and dry pastoral properties receiving less than 200 millimetres per year. However, most of these enterprises can be described as variations of five principal methods of production:

- self-replacing flock
- prime lamb production
- First Cross ewe breeding
- woolgrower operation
- fattening.

Self-replacing flock

This type of production system is the most common type of flock used and is the main one used in the Merino industry. Commercial flocks of Polwarth and Corriedale sheep and all studs also use this breeding program.

The principle feature of this production method is selecting superior young replacement ewes from the previous year's flock and returning them to the main ewe flock. The oldest age group of ewes are commonly referred to as cast for age (CFA) and are culled each year.

The flock age structure for this type of system will depend on productive life of the ewe, mortality rates and weaning percentages. A typical flock structure is shown in Figure 1.8. In this case, there are 1000 ewes in the

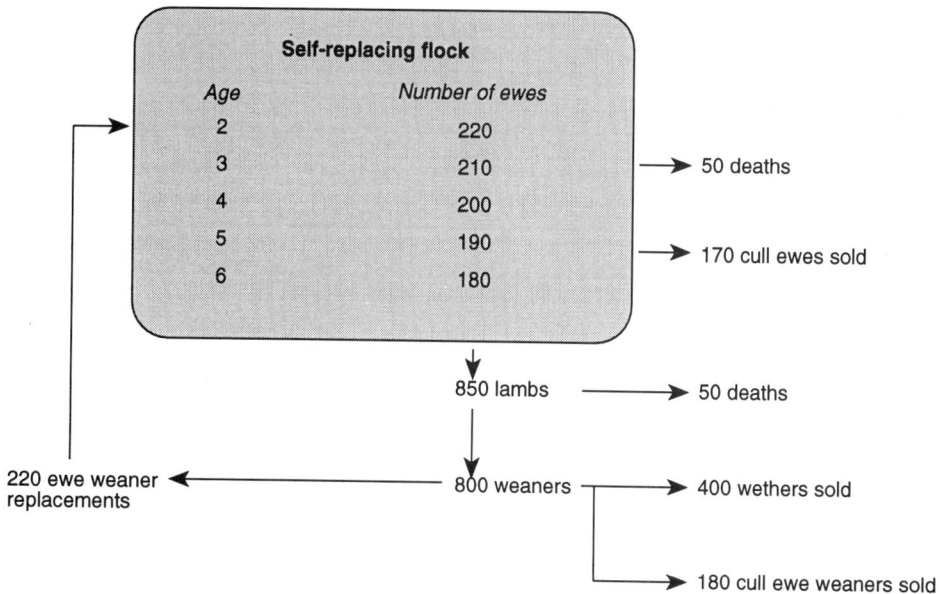

Figure 1.8 *Self-replacing flock of 1000 head, showing losses and replacements*

15

flock, the productive life is five years, death rate in the ewes is 5.0 per cent and the weaning percentage is 80 per cent.

First Cross ewe breeding

The production of First Cross ewes for prime lamb mothers uses a flock of Merino ewes joined to a Border Leicester ram. The ewe portion is retained on the property for breeding prime lambs or sold to specialist prime lamb producers. The wether portion is sold to the meat trade as later maturing heavy lambs.

In this example, Figure 1.9, of a First Cross ewe breeding flock, the Merino ewe portion of the flock would be maintained by purchase or from another breeding enterprise on the property. Weaning percentage is 80 per cent and mortality rate in the ewes is 5.0 per cent. Note that many producers join their older ewes to a Border Leicester ram to breed First Cross ewes.

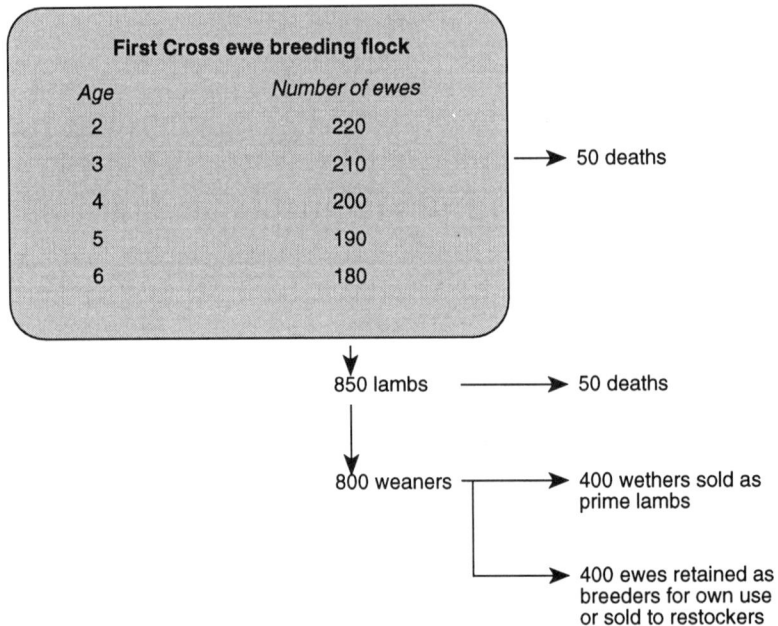

First Cross ewe breeding flock

Age	Number of ewes	
2	220	
3	210	→ 50 deaths
4	200	
5	190	
6	180	

850 lambs → 50 deaths

800 weaners → 400 wethers sold as prime lambs

→ 400 ewes retained as breeders for own use or sold to restockers

Figure 1.9 *First Cross ewe production, from a flock of 1000*

Prime lamb production

The prime lamb industry predominantly uses First Cross ewes as mothers and a wide range of terminal sires to produce carcases suited to different weight and fat requirements of the meat industry. Some prime lambs are also produced using Merino or other pure breeds as mothers.

In Figure 1.10, the ewe flock is 1000, weaning percentage is 120 per cent and death rate of the ewes is 5.0 per cent.

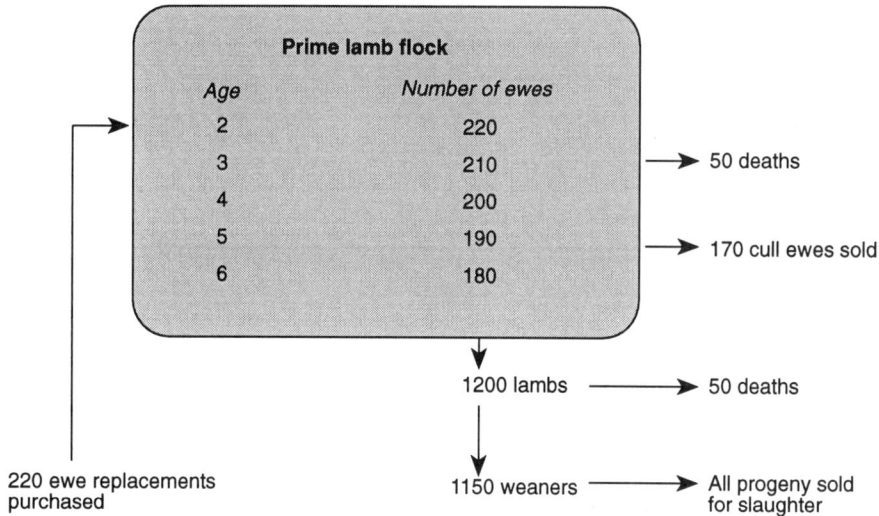

Prime lamb flock

Age	Number of ewes
2	220
3	210
4	200
5	190
6	180

→ 50 deaths

→ 170 cull ewes sold

1200 lambs → 50 deaths

1150 weaners → All progeny sold for slaughter

220 ewe replacements purchased

Figure 1.10 *Prime lamb flock*

Wool growing operation

A wool growing enterprise consists of a Merino wether flock. These sheep are purchased usually as weaners or young sheep. The wethers are run until wool quantity and quality begin to decline. They are then sold as older, fat wethers to the local or export meat trade or for live wether exports.

In the example in Figure 1.11 of a wether only flock, the mortality rate is 5.0 per cent.

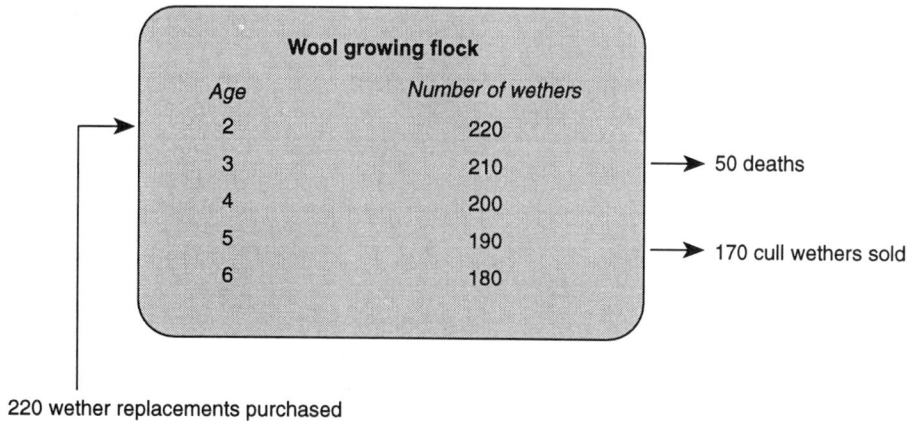

Wool growing flock

Age	Number of wethers
2	220
3	210
4	200
5	190
6	180

→ 50 deaths

→ 170 cull wethers sold

220 wether replacements purchased

Figure 1.11 *Wool growing flock*

17

Fattening

This group of producers are involved in other enterprises and only fatten stock at certain times of the year to fit in with these enterprises. For example, an irrigated lucerne hay producer may fatten lambs for three months of the year after the last cut of hay in autumn.

Some producers on irrigated properties fatten stock as their main activity. The stock selected for fattening will depend on availability and expected profit levels. Stock used in this enterprise may range from fattening young store lambs to cast for age ewes and wethers.

Sources of rams

Rams purchased by breeders come from either parent, daughter or general studs. A parent stud is a long established and often larger stud while a daughter stud is one that has developed by purchasing ewes and all rams from a parent stud. A true daughter stud only purchases rams from the one source. Many studs now function as a general stud, taking rams from several sources that will breed the type of animal they wish.

Annual sales are held on most studs where a selection of the better rams is presented. The remainder of the year's production of rams is graded and offered for private sale. The top few rams are retained each year to become stud sires.

There have been several groups of producers who have started group breeding schemes in an attempt to breed a more suitable Merino for their area. Group breeding schemes work by identifying the superior ewes in each flock and combining these animals into a nucleus flock. Rams are bred from this flock and distributed back to each member.

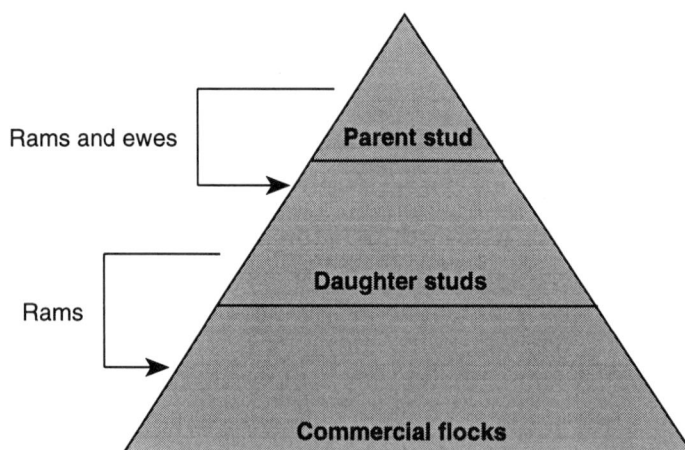

Figure 1.12 *The traditional structure of the Merino industry*

Production areas

There are three sheep production areas in Australia. These are the high rainfall area, the sheep-wheat area and the pastoral area (Figure 1.13). The production and sheep management principles used within each area follow the same basic rules even though there is considerable variation in country and sheep types used. In addition sheep may be kept in intensive production.

High rainfall area

The high rainfall area is restricted to the tablelands of New South Wales and Victoria, parts of Tasmania and the south-western corner of Western Australia. The high rainfall zone covers about 5.0 per cent of the sheep growing area but contains 20 per cent of total sheep numbers. Rainfall in these areas varies from 700 to 1000 millimetres per annum. Temperatures are cold in winter with occasional snow falls and cool to warm in summer.

The combination of good moisture levels and cool temperatures means there are large areas of established improved pastures. Pasture production is lowest in winter due to the cold temperatures and pasture quality and quantity are at their peak from mid spring to mid summer. Stocking rates are dependent on the extent of pasture improvement carried out and vary from 2 dry sheep equivalents (DSE) per hectare on unimproved native pastures to in excess of 12 DSE per hectare.

Self-replacing flocks, prime lambs and wether enterprises are situated in this area.

The main sheep breed used in this area for both breeding and wool production is the superfine and fine wooled Merino. This sheep type is used in the high rainfall areas because it is very resistant to fleece rot (and so flystrike). Full value is paid for the fine wools produced from this area because of very low levels of vegetable matter and dust penetration into the wool staple. There is some use of medium wool Merinos due to the higher fleece weights attained. When these types are used, there is considerable emphasis placed on resistance to fleece rot andstylish wool (bright, crimped clean wool) when breeding or buying medium wool Merinos.

Specialist prime lamb producers purchase First Cross ewes from the sheep-wheat zone and select a suitable prime lamb sire to meet market requirements.

Many fine and superfine Merino and British breed studs are located in the high rainfall area.

More time is spent in managing a flock in the high rainfall zone compared to the other areas. Cropping is not an option on most properties except for pasture improvement or sowing an area of fodder crop like oats and so grazing sheep may be the only enterprise. It is essential to carry out all the husbandry treatments necessary to maintain the health and production levels

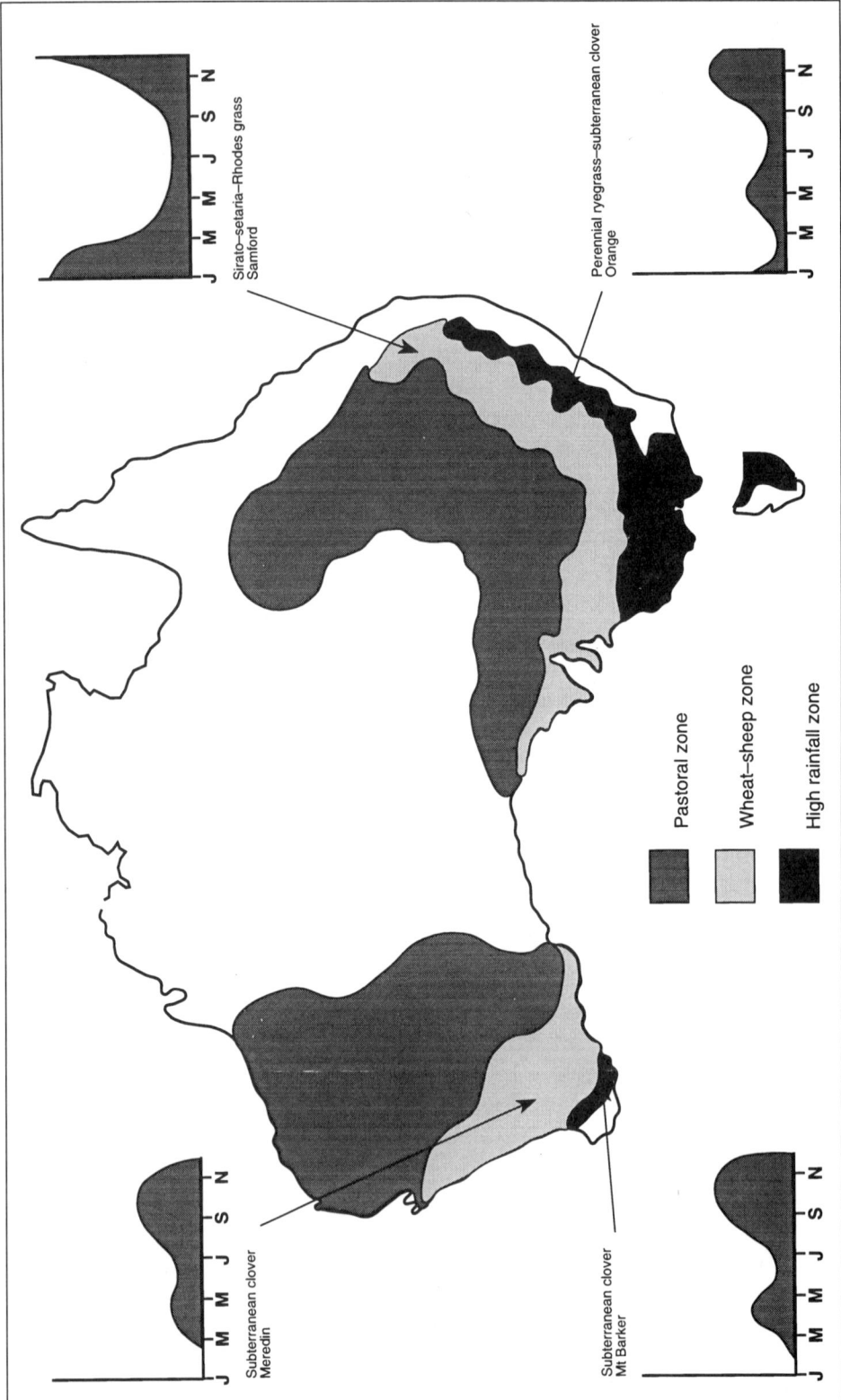

Figure 1.13 *Sheep production areas, showing pasture growth curves of four regional centres* *(Source:* Chapman *et. al.* 1973)

of a flock at higher stocking rates and to maximise production to ensure economic survival on smaller holdings.

Sheep-wheat area

The sheep-wheat zone is situated between the high rainfall and pastoral zones throughout Australia. This is a mixed farming area and sheep production is combined with growing crops and grazing cattle.

The sheep-wheat zone contains 60 per cent of total sheep numbers in Australia. Rainfall in these areas varies from 400 to 650 millimetres per annum. Temperatures are mild in winter with frosts in some areas, and warm to moderately hot in summer. Stocking rates depend largely on rainfall and varies from 1 DSE per hectare on the drier fringe to over 10 DSE per hectare in the better rainfall areas with pasture rotations.

There are some areas of established improved pastures. Pasture production is lowest in winter due to the colder temperatures. Pasture quality and quantity are usually good in autumn after the first significant falls of rain and at their peak from early to late spring. Fodder crops are used to supplement periods of low pasture production in autumn and winter.

Self-replacing flocks, First Cross ewe production, prime lambs and wether enterprises are situated in this area. The main sheep breed used for both breeding and wool production is the medium wool Merino. There are higher levels of vegetable matter and dust penetration because of fallowed (ploughed) country and the harsher, drier conditions, and considerable emphasis is placed on selecting sheep able to withstand this.

Prime lamb producers are located in the wetter parts of this region or where there is an opportunity for irrigation. First Cross ewes are produced throughout this region for sale to specialist prime lamb producers. There are also many medium wool Merino studs in this zone.

Sheep management in this zone is often a balance between farming operations and a beef cattle enterprise. Essential husbandry operations still need to be carried out but timing is affected by the work requirements of other enterprises. Decisions on where to run different classes of sheep are made taking into account the need to graze fallow paddocks, the needs of another enterprises, as well as the nutritional requirements of the sheep.

Pastoral

The pastoral zone covers the main portion of sheep growing areas in Australia. It is a harsh, dry environment with rainfall from 150 to 350 millimetres per year. Pastures are native grasses and herbages with very little pasture improvement. Stocking rates vary from one sheep requiring between 2 and 8 hectares.

Sheep types are predominantly large frame medium and strong wool sheep run as a self-replacing flock. This type of animal is preferred because of their robust constitution, ability to walk long distances to forage, and easy-care conformation. There are some wether only enterprises but most wethers are run in association with a breeding flock.

Sheep management practices are very different from the other zones. Sheep are yarded only for shearing and lamb marking on some properties. In many cases, sheep may not be crutched. Mustering sheep in this zone is a major cost because of the distances involved and the occurrence of scrub or woody weeds that make mustering very difficult. Sheep are often mustered using light planes or helicopters to co-ordinate on ground activities.

Feedlots

Feedlots are a means of finishing stock for sale. The stock are kept in a confined area to limit their movement and fed high protein, high energy feed mixes This aims to achieve efficient conversion of feed to gain the highest possible growth rates. Generally all replacement stock and fodder are purchased in, and so profit margins need to be ensured prior to the commencement of a feeding program. In larger feedlots, stock are forward sold to ensure the profit margin.

Sharlea

This enterprise involves the shedding of fine and superfine wool Merino wethers for the production of high quality clean wool. The sheep are fed a grain and hay ration and kept indoors throughout the year.

The main advantages of this operation are:

1. It is a completely controlled environment ensuring even wool growth.
2. Wool is kept free of dust and vegetable matter.
3. Wool of a consistent micron range and tensile strength can always be produced.

Sharlea wool takes dye differently to other wools so that a brighter coloured fabric is attainable.

CHAPTER 2

Flock management

Ewe management

Fertility of the flock, measured as the number of lambs weaned per 100 ewes joined, is a key factor in running a profitable breeding enterprise. Regularly high weaning percentages can only be achieved by careful management of the ewe flock.

Physiology of the ewe

Oestrus

The oestrus cycle of the breeding ewe is a recurring period of 17 days and is controlled by levels of hormones in the bloodstream. For approximately 14 days of this cycle there is minimal hormonal activity and levels remain constant. At about Day 15, falling progesterone levels affect the ovary and a new egg begins to develop. Rising oestrogen levels cause the ewe to come on heat (show behavioural oestrus where the ewe will actively seek out the ram and stand to be served). The ewe then ovulates (releases the egg into the reproductive tract) in readiness for fertilisation. Adult ewes in good condition will remain on heat for a period of 24 to 36 hours but this is reduced in maiden ewes or ewes under nutritional stress (Figure 2.1).

Control of the oestrus cycle is necessary in artificial breeding programs to be able to inseminate all ewes at the same time. This is usually done using progesterone implants, sponges or plastic release devices impregnated with the hormone progesterone which are inserted carefully into the vagina. The hormone is absorbed by the ewe and when the implant is removed, most of the ewes will cycle at the same time.

Ovulation rate

The ovulation rate is the number of eggs released per 100 ewes joined. The ovulation rate is important because this will determine the number of twins

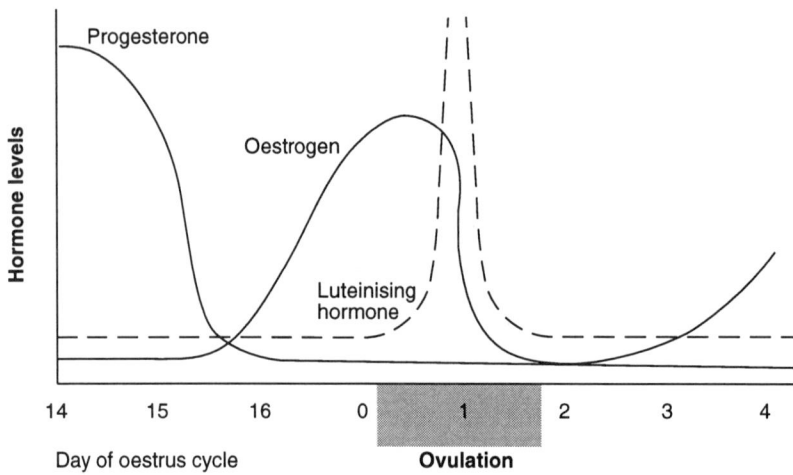

Figure 2.1 *The oestrus cycle*

born and the overall lambing percentage. Ovulation rate is mainly determined by the breed of sheep, the condition of the ewe at joining, nutrition levels at joining and the time of the year.

Table 2.1 *Breeding table showing gestation periods*

Mating date		Lambing date		Mating date		Lambing date		Mating date		Lambing date	
Jan	1	May	30	May	7	Oct	3	Sept	10	Feb	6
	8	June	6		14		10		17		13
	15		13		21		17		24	20	
	22		20		28		24	Oct	1		27
	29		27	June	4		31		8	Mar	6
Feb	5	July	4		11	Nov	7		15		13
	12		11		18		14		22		20
	19		18		25		21		29		27
	26		25	July	2		28	Nov	5	Apr	3
Mar	5	Aug	1		9	Dec	5		12		10
	12		8		16		12		19		17
	19		15		23		19		26		24
	26		22		30		26	Dec	3	May	1
Apr	2		29	Aug	6	Jan	2		10		8
	9	Sept	5		13		9		17		15
	16		12		20		16		24		22
	23		19		27		23		31		29
	30		26	Sept	3		30				

Gestation

Gestation is the period which commences with fertilisation of the egg and finishes when the ewe lambs. The gestation period of the ewe is 147 to 155 days (Table 2.1).

Pre-joining management

Management of the ewe flock is primarily aimed at having the ewes in good condition and free from any health problems at joining. The length of the pre-joining period will vary depending on the length of joining the previous year. A six week joining allows for a period of about 12 weeks for the ewe to increase body weight following weaning, prior to the next joining.

Nutrition

Body weight targets at joining will vary between ewes of different frame. The critical ewe body weights in Table 2.2 below are the weights (free of fleece) below which the ewe flock will have low and unpredictable lambing results.

Table 2.2 *Pre-joining body weight for ewes*

Breed/type	Critical ewe body weight (kg)	Desired ewe body weight (kg)
Merino — fine wool	37	45
Merino — medium wool	40	50
Merino — strong wool	43	55
Crossbreed	45	55

Condition scoring sheep when scales are not available is an excellent method of assessing and monitoring ewe condition. (See page 112.) It is essential to yard the sheep and condition score a reasonable sample from the mob. Sheep can not be condition scored from a motorbike or the front of a ute! Do not condition score the first or last of the mob into the yards as they may not provide a fair estimate of the bulk of the mob.

Ewes should be in at least a condition score 3 at joining and condition score 4 is better. Feeding levels should be sufficient to achieve these levels by joining. When dry or drought conditions exist, the breeding flock will need to be fed supplements to make up for any shortfall in pasture quantity or quality. Suitable supplements to increase pre-joining body weight include good quality hay, oats and lupins.

Husbandry

The pre-joining period is an excellent time to carry out husbandry operations. Heavily pregnant ewes are harder to handle and health problems are possible if the ewes are yarded for long periods. Pre-joining mob sizes can be significantly larger than during joining or lambing, which reduces time in driving several mobs to the yards. The actual procedures required will depend on the production area and management program of the property, (see Chapter 8, Husbandry operations).

Management during pregnancy

Nutrition

Management of the pregnant ewe for the first three months of pregnancy is little different to management of dry sheep on the property. The nutritional needs of the ewe are very similar to those of a dry sheep. Body weight should be held at joining levels or allowed to increase slightly. Ewes should not be allowed to become over-fat by mid-pregnancy. Ewes in over-fat condition at lambing will produce larger lambs and have a higher rate of difficult births.

During the last eight weeks of pregnancy, the nutritional needs of the ewe increase. The lamb foetus rapidly increases from about 500 grams at three months to approximately 4 kilograms at birth. The foetus gains between 80 and 90 per cent of its weight during this period.

If the ewe does not put on at least 4 kilograms in body weight (the weight of the lamb) during this period, she is actually losing condition. Ewes carrying a single lamb require double the amount of feed to allow for lamb growth compared to a dry sheep.

Pasture conditions and feeding levels that fall short of the ewe's requirements result in lambs that will be lighter at birth, be weaker and have a lower survival rate. If feed intake is much lower than the ewe's requirements in the last four weeks of pregnancy, the onset of milk production in the ewe will be delayed, adding to lamb deaths. Under these conditions, a significant number of ewes may be affected with pregnancy toxaemia. This is a metabolic disease caused when the ewe's body tissues are broken down to provide energy and protein for lamb growth.

Ewes placed on excessive feed may produce large lambs. Large lambs are more likely to have difficult births and to suffer some type of birth injury.

Husbandry operations

Ewes must be kept in good health in late pregnancy. Any problem like foot abscess, footrot, flystrike or a broken mouth will reduce grazing ability and feed intake.

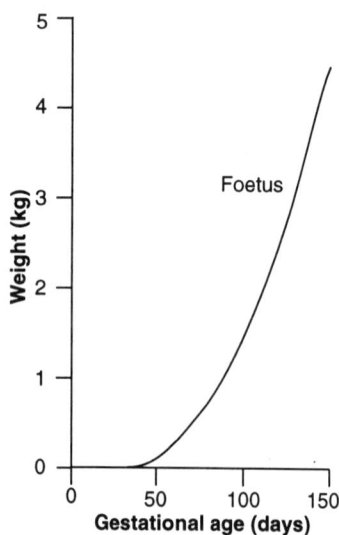

Figure 2.2 *Typical pattern of foetal growth*

Many producers drench and vaccinate their ewe flock within two weeks of lambing and move the flock to a fresh paddock. However, extra care must be taken performing any husbandry operations when the ewe is heavily pregnant and operations like foot trimming that turn the ewe over for an extended period should be avoided. Ewes should only be yarded for the minimum time required to perform the operation.

The management year should be planned to avoid shearing ewes on the point of lambing. Locking ewes up overnight can result in animals being affected by pregnancy toxaemia, and cold conditions after shearing add to already high nutritional needs.

Pregnancy testing

Pregnancy testing (or scanning) can be carried out using ultra-sound. Most sheep areas have several contractors offering this service. Ewes can be scanned to determine if they are pregnant or empty and an additional service is available to determine if the ewe is carrying single or twin lambs.

Scanning allows for:

1. Division of the flock into single and twin bearing ewes to allow for the higher feed requirements of twin bearers.
2. Division of the flock at lambing to allow closer supervision of twin bearing ewes if required.
3. Early culling or reduced feeding levels of ewes failing to conceive.

Ram management

Physiology of the ram

Rams are joined to a breeding flock at a low percentage, typically 2.0 per cent or 1 ram to 50 ewes. A ram needs to be able to serve many ewes in a short period of time and achieve a high conception rate. To do this, the ram must produce a sufficient volume of good quality semen, have a high libido (sex drive) and be physically sound.

Semen production

Semen is produced in the testes and mature in the epididymis over a seven week period. Semen production occurs at a temperature several degrees lower than normal body temperature. The temperature of the testes is regulated by sweat glands and the distance the testes are held from the body. Any factors that cause the body temperature of the ram to rise including disease and high daily temperatures may cause temporary infertility.

General management

The rams should be moved slowly in the cool part of the day. Some producers truck rams to the paddock where longer distances are involved to avoid any heat stress.

Rams should have access to good feed, water and shade. General health should be checked regularly, especially flystrike around the poll, foot abscess, footrot and lice. The drenching program used for the rams should be in line with the ewe flock and all vaccinations completed and up to date. Rams will go to each flock of ewes on the property, and any disease or parasite problem that exists in the ram flock can be quickly spread over the property through infected rams.

Pre-joining

Pre-joining management of rams is critical to fertility levels during joining. The aim is to have rams in peak physical condition and able to produce healthy, viable semen for the six week joining period.

A physical examination of all rams at least ten weeks before joining should be made. This allows time for replacement rams to be purchased and prepared for joining if rams are culled. The main areas to be examined are the toes, tossel, testes and teeth — the four Ts.

Toes
The feet should be checked for:
- Overgrown horn that may make walking difficult. Trim if necessary.
- Footrot or shelly hoof problems, indicated by any separation between the junction of the hard and soft part of the foot.
- Foot abscess, abnormal swelling between the hard horn of the hoof and leg.

Tossel
This is the area around the penis. The penis is checked for any abnormalities, injury or infections that will affect mating with the ewe. The sheath may be checked for physical damage from grass seed or shearing cuts.

Testes
To check the testes, they can be carefully palpated.
- The testes should be of even size, firm and be large enough to ensure adequate semen production.
- The scrotum is checked for abnormal thickening of the skin or injuries.
- The epididymis is carefully felt along the head and tail for any swellings that may be caused by disease or physical damage.
- Spermatic cords are felt for any lumps or swellings that may prevent sperm movement.

Any swellings in the testes or epididymis will reduce fertility and swellings on both sides will probably render the animal infertile. Swellings may be associated with brucellosis (a reproductive disease of sheep) and if this is suspected a vet should be called to check out the ram flock.

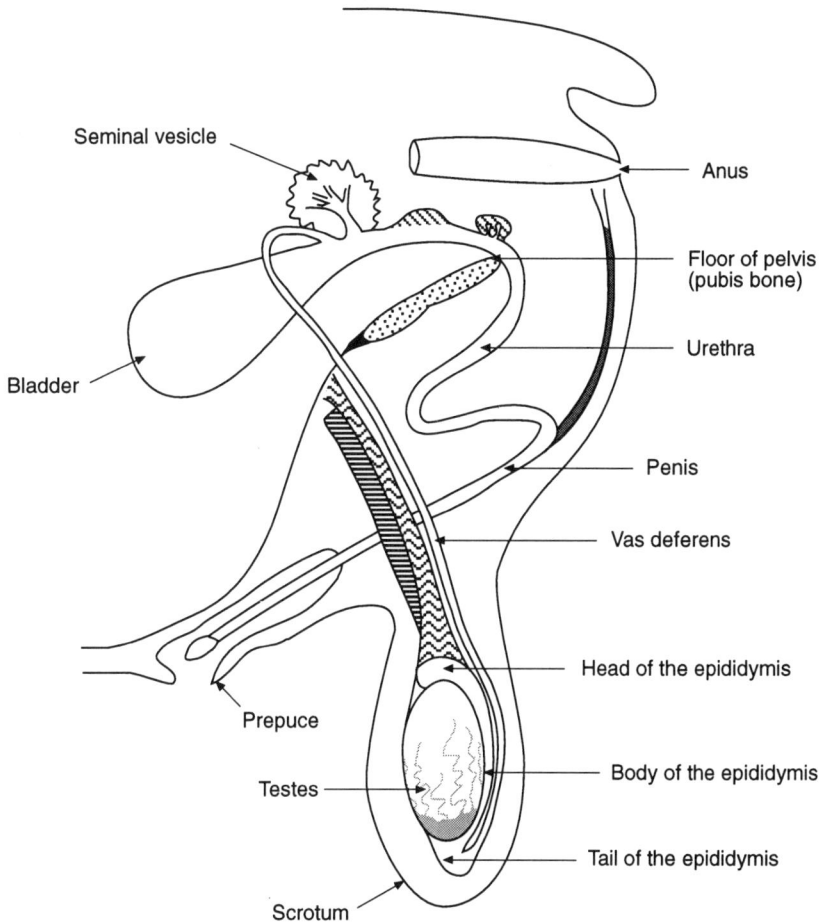

Figure 2.3 *Ram's reproductive organs*

Teeth

The mouth is checked:

- to determine age,
- to make sure the mouth is sound so the ram can feed effectively,
- to ensure teeth meet the hard palate evenly.

Nutrition

Rams should be on good quality feed and in condition score 3 to 3.5. Rams should not be allowed to become over-fat. Fertility is less in over-fat rams, especially in hot conditions, and serving capacity will be reduced.

The size of the testes and semen production can be altered by the quality and quantity of feed available before joining. Feeding high energy and high protein supplements for a short period of six to eight weeks prior to joining will result in increased semen production. However, this feeding strategy should not be used to compensate for neglect and poor feeding levels for the remainder of the year.

Where feed quality has been poor due to dry or drought conditions, the feeding of a high protein supplement such as lupins gives excellent results. In this situation, the feeding of some green supplement or good quality lucerne hay can provide vitamin A that may be deficient.

Shearing

Rams should be shorn twice a year, and ideally joined carrying two to three months' wool. However, rams should not be shorn and dipped within six weeks of joining as the stress involved with shearing and dipping can reduce fertility. Less than six weeks' wool will increase body temperatures and so reduce fertility.

CHAPTER 3

Selecting stock for breeding

Classing and selecting sheep is an important part of preparing a flock for joining, preparing stock for sale to a restocker or purchasing rams from a stud. Selecting sheep can be visual or by the use of production information or a combination of both.

Genetic principles

The aim of any selection process is to pick out the most desirable animals for breeding purposes to improve the overall quality and profitability of the flock. Before we discuss how to do this, there are a few genetic principles we need to understand.

Heritability

This is the proportion of the parents' superiority passed onto the offspring. If a ram is an outstanding individual for fleece weight, only a percentage

Table 3.1 *Heritability of various sheep features*

Trait	Heritability (%)
Greasy fleece weight	30–45
Clean fleece weight	28–45
Staple length	31–65
Fibre diameter	37–57
Weaning weight	20–25
Hogget weight	35–60
Lambs born/ewe joined	3–20
Lambs weaned/ewe joined	3–15
Susceptibility to fly strike	35–40

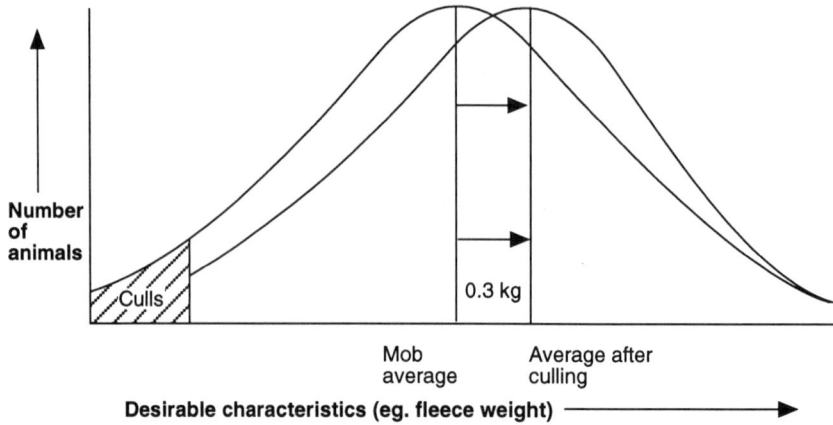

Figure 3.1 *Bell curve for Flock 1, showing the effect of light culling*

will be passed on to the lamb. The heritability for fleece weight is about 40 per cent. A heritability of 40 to 50 per cent is very high, and 20 per cent is very low.

Selection difference

This is how much better the selected animals are compared to the average of the flock before culling. The smaller number of better animals selected, the higher will be the selection difference. The spread of animal performance across the flock may be graphed in a bell curve, as in Figure 3.1

In this case only a small portion of the flock has been culled. The animals selected for breeding have a small advantage over the unclassed flock. The selection difference in this case is quite low.

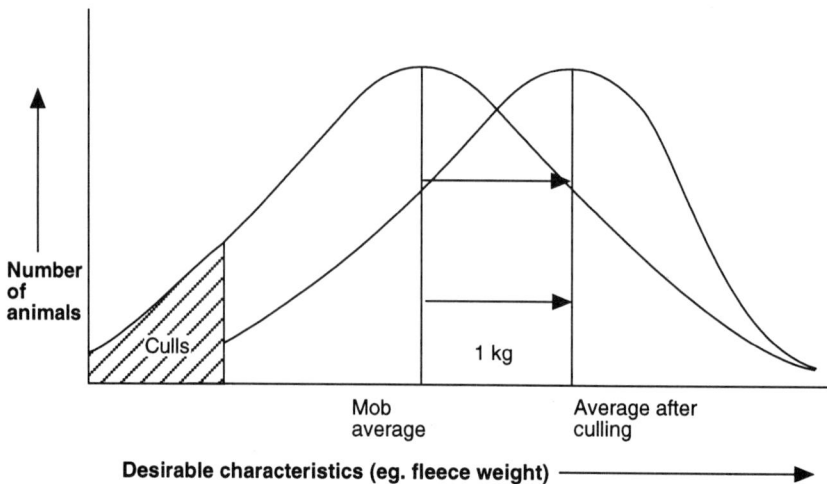

Figure 3.2 *Bell curve for Flock 2 with a heavier cull and greater selection difference*

In Flock 2 (Figure 3.2) a much greater proportion of the mob has been culled. The selected animals have a large advantage over the previously unclassed flock, and there is a difference made to the flock average, which is quite high.

Selection difference may be measured in units appropriate to the criteria used for culling. For instance, if Flock 1 was graded according to wool production, and culling carried out to increase the average fleece weight, this course of action may be expressed as imposing a selection difference of say 0.3 kilograms.

Genetic gain

This is the advantage gained in each generation from selecting the better animals.

Genetic gain is found by using the following formula:

Heritability × selection difference = genetic gain

In Flock 1, the genetic gain per generation would be:

0.3 kg (selection difference) × 40% (heritability) = 0.12 kg wool cut

In Flock 2, the genetic gain per generation would be:

1 kg × 40% = 0.4 kg wool cut

These genetic principles apply to visually classed characteristics as well as production information given as an example here. It is important to remember:

- Not all the advantages of the sire or dam will be passed on.
- Different characteristics have different heritabilities.
- The more low producing or undesirable animals culled, the better will be the remaining flock, and the better will be their lambs.

Visual classing

Sheep are classed in a handling race or a classing box that allows the entire animal to be seen and partially restrained so the wool and mouth of each sheep can be inspected. When preparing to class sheep visually, do not run a mob of sheep into the yards and up the race and start classing. Often the weaker and less suitable sheep will be toward the rear of the mob. Walk through the mob and have a careful look at the entire group first, and go through a couple of races of sheep before making any culling decisions. This gives a better impression of the overall mob and the types of conformation and wool faults that may need to be removed.

Do not look for any slight fault in a sheep so it can be culled — many productive and profitable sheep will have some small problem. Only cull sheep where the fault is a very serious one or where the animal has several faults of a lesser degree that combined together make it unsuitable for the mob.

Sheep may be culled for conformation or body faults, and wool faults.

Conformation faults

These conformation faults apply equally to sheep run for wool or meat production.

Mouth

The jaws should meet evenly to allow the sheep to feed efficiently.

Wool coverage

The amount of wool around the head and down the legs varies with different strains of sheep. Generally sheep with excessively "muffled" or woolly heads are culled, especially if wool is present in front of the eyes. These sheep will become "wool blind" and are far more likely to get grass seeds in their eyes.

Pasterns

The pastern is the lowest joint of the leg. The angle formed between the hoof and the leg must not be weak or sloping. This reduces the walking ability of ewes and is a particularly bad fault in rams. Weak pasterns will reduce serving capacity, since a mounting ram places all his weight on the hind legs.

Hocks

"Hocky" sheep have their knees too close together. They are usually culled from a mob because of appearance and probably reduced walking ability. Hocky sheep have problems with urine and dung stain down the back legs and increased incidence of fly strike.

Shoulders and back

All animals with poor back and shoulder conformation should be culled. Faults in this area mean the animal is more likely to develop fleece rot and blowfly strike. "Devil's grip" is one particular conformation weakness seen as a distinct depression just behind the shoulders.

Body size

Small animals are culled. This should not be carried out before the animals are 14 months of age, becuse the smaller sheep at younger ages may be twins or the younger portion of the drop.

Teat damage
Ewes with damaged teats should be culled because they are not able to feed their lamb effectively. Teat injuries usually occur at shearing and every effort should be made to avoid them.

Wool faults

Pigmented wool fibres
All sheep with black spots should be culled from the mob. Occasional black fibres are present in all Merino sheep but the levels of these are much higher in sheep with black areas. There may be some genetic relationship between sheep with excessive areas of black and brown spotting on the non-wool areas of the body around the head and legs.

Hair fibres
Hair fibres are a fault and Merino sheep showing hair should be removed. Hair fibres are much coarser than wool fibres and are usually found on the breech. A small number of sheep will show a "halo" effect with a fine covering of hair over the entire body.

Fleece rot
Fleece rot areas occur along the back and is important because it reduces the value of wool and makes the sheep more likely to get flystrike. This is one area of sheep selection that requires a careful approach. In very wet years, a large portion of the mob may show some fleece rot so culling levels should reflect the current season.

Wool outside flock parameters
Consider culling sheep with a wool type obviously very different from the remainder of the mob. As an example, in any drop of sheep there will be a small number of sheep with much stronger or shorter wool than the mob average. Removing these at sheep classing time reduces "off-type" fleeces at shearing.

Measuring production

There are several aids to selection that are available which make use of measurements. These may measure the production levels of individual sheep, progeny of different sires or production levels of different bloodlines. All animals that are measured for comparison and selection (or compared visually) must be treated exactly the same and run together as a mob. By doing this, the differences observed or measured are differences due to genetic differences betweed the sheep, not differences due to better nutrition or management.

Wool production traits

Greasy fleece weight
This is measured by weighing shorn fleeces. Greasy fleece weight information is usually given as an index to make the information easier to interpret. The average of the mob is given a value of 100 and the rest of the mob compared to this figure. A value of 120 means the sheep has a fleece weight 20 per cent (120 – 100) above average, 125 is 25 per cent above average and 90 is 10 per cent below average.

Micron deviation
Wool thickness is measured in microns, or millionths of a metre. It is measured by sending a mid-side sample from the animal to a wool testing service. Micron is compared to the average of the mob and given as a plus or minus figure. A sheep with a micron deviation of + 1.5 is 1.5 microns stronger than the average of the mob.

Clean fleece weight
This information comes from greasy fleece weight adjusted for yield. Yield is the amount of fibre without dust, grease or vegetable matter and is measured by a wool test service. Clean fleece weight figures for individual animals can be presented as for greasy fleece weight.

Body weight
Body weights are measured usually after shearing and like fleece weight information are provided as an index where the mob average is given the value of 100. Measured information may be combined into a selection index or a single value that assists the breeder in identifying the more productive animals in the flock or replacement rams that are most suitable.

Wether trials

Wether trials are conducted by interested producers in an area by running a group of wethers, representing many different bloodlines of sheep, together for several years. Production levels are measured each year and each group of wethers compared. Combined information from several hundred of these trials is now available. This information is a useful indication of the relative production levels of different bloodlines of sheep, especially fleece weights and mircon comparisons.

Meat production traits

Objective measurements of meat production traits are provided by many breeders of terminal sires. This allows prime lamb producers to select sires that have good growth rates, but still produce lambs with the desired level of fat coverage.

There are two measurements provided. The firdt is the growth percentage, measured as the growth rate of individual animals compared to the remainder of the drop. This figure is presented as an index, as for fleece and body weight measurements.

The second measurement is the fat class of individual rams, which is given as the difference up or down from the average of the ram drop.

CHAPTER 4

Breeding

Joining

Deciding on when and how to join a breeding flock of ewes will affect all other aspects of the management of that flock. Lambing percentages, the survival of lambs to weaning, and wool quantity and quality will all be determined to some extent by joining.

When to join

The time selected for joining will depend on many factors and will affect all aspects of sheep management over the year. Sheep are seasonal breeders: decreasing daylight will cause more ewes to ovulate, they will ovulate at a higher rate and come on heat stronger for longer periods. This is any time after 22 December and peaks in mid autumn. Ewes will still join successfully outside this time, especially if they are in good condition and gaining weight. Merinos will breed outside the optimum time better than will British breeds and their crosses.

Other factors that may affect the time of joining:

- Lamb survival will be higher when lambing at the same time as neighbours because predators will be spread over a larger area.
- Climatic conditions at the planned time of lambing should be taken into consideration.
- Target markets should be considered so that stock are prime when markets are at their peak.
- Labour availability; on mixed farming properties lambing needs to be planned to avoid clashing with other major operations.
- Joining is timed to give a lambing date when good levels of pasture are expected to be available.

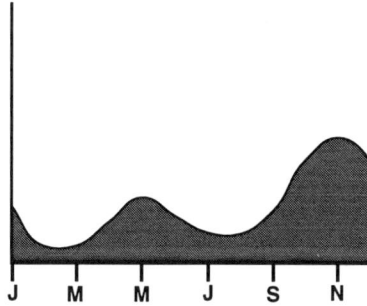

Figure 4.1 *Pasture growth curve for Orange, NSW. Note the dip in July/August but the rise thereafter*

Peak feed demand of a breeding flock will occur in the last month of pregnancy and the first two months of lactation. A heavily pregnant ewe requires double the normal amount of feed, and a lactating ewe requires 2.5 times maintenance requirements. Good pasture quality and quantity is also required to finish lambs for slaughter or grow out Merino lambs after weaning. It is important to take these requirements into consideration.

If lambing was planned for July/August in the Orange area of NS W, there would be an expected cost for supplementary feeding (Figure 4.1). However, once past this period, there would be high quality pasture available until December. An earlier lambing date would be more suitable in a climate with an expected flush of feed in autumn and a fall in in late spring.

Length of joining

A ewe will come on heat and ovulate every 17 days during the breeding season. A joining period of six weeks (42 days) will allow every ewe to come on heat at least twice during joining and about half will have three cycles. If the joining period is extended to 52 days, all ewes will have three cycles to get in lamb.

There are two advantages in a six to eight weeks joining period:

1. The lamb drop will be more even and there is less variation in age. This makes the marketing of the progeny easier and the classing of replacement stock more accurate.
2. Keeping the joining period shorter allows maximum time for the ewe to regain body weight before joining the next year.

Table 4.1 *A ewe's breeding year with a joining period of six and eight weeks*

Year 1 Joining	Pregnant or rearing a lamb	Dry ewe: recovering condition	Year 2 Joining
6 weeks	34 weeks	12 weeks	
10 weeks	34 weeks	8 weeks	

If the joining period is increased to ten weeks to allow lightly conditioned ewes to get in lamb, or to solve other infertility problems in the flock, then the recovery time will suffer. This often results in lower condition at the second joining, reduced ovulation rates and lower lambing percentages.

Joining percentages

The joining percentage is the average number of ewes allotted per ram It will vary widely throughout the industry.

When planning ram requirements it is useful to think in terms of the ram/ewe contact. The ewe will actively seek out a ram as long as one is visible when she comes on heat. A suitable joining percentage would be higher where there are factors like topography, scrub or large paddocks discouraging close and continuous ram/ewe contact. Where paddocks are smaller, cleared and with only one watering point available to the, stock, a high joining percentage (say 3.0 per cent) can lead to excessive competition between rams.

A general recommendation under normal conditions is a joining percentage of 1 per cent, plus one extra ram. For example, 800 four year old ewes on good feed in relatively small paddocks with experienced rams would be joined with eight rams (1.0 per cent) plus one, giving a total of nine rams. There are potential problems with this approach where ewe numbers are small. For instance failures may result from joining a small flock to one or two rams which have fertility problems.

Where conditions for ram/ewe contact are not ideal, or some other factor to successful mating is present, two extra rams are added for each factor. These factors may include:

- Poor nutrition levels.
- Low levels of feed causing the flock to disperse widely.
- Joining outside the breeding season.
- Joining maiden ewes.
- Using a large proportion of inexperienced rams.
- Large paddocks, especially with multiple watering points.
- Hilly or scrub covered paddocks.
- High temperatures that affect semen quality.

For example, joining a maiden ewe flock of 400 in a large, hilly paddock:

1.0 per cent	= 4 rams
+1	= 1 ram
maidens	= 2 rams
large paddock	= 2 rams
hilly terrain	= 2 rams

giving a total requirement of 9 rams.

Figure 4.2 *Ram fitted with harness*

Ram harnesses

A ram harness is used to monitor joining activity in a flock, to draft ewes into early and late lambing groups, or to establish if an individual ram is working. The harness straps onto the shoulder area and holds a replaceable crayon against the brisket. When the ram mounts and serves a ewe a crayon mark is left on the ewe's rump. The crayons can be obtained in three different heat grades to suit all conditions. The fit of the harness needs to be monitored on a regular basis over the joining period and tightened as the ram loses weight.

Types of joining

There are two ways in which a flock of ewes can be joined. These are syndicate joining, or single sire joining.

A syndicate joining uses a group of rams to join to a flock of ewes. The main advantages are that a lower performing or infertile ram in the syndicate will be compensated for by other rams, and that only one paddock is needed. This method is the most common and is used for 98 per cent of joining in the Australian sheep industry.

A single sire mating joins one ram to a small flock of ewes. The main advantage of this method is the reproductive performance of the ram and the production levels of his progeny can subsequently be accurately determined. For this reason, many stud sires are joined using this mating method. The disadvantages are that if the ram has low fertility, the ewes will have a poor lambing and more paddocks are needed to join many smaller mobs of ewes.

Lambing management

The final profitability of any breeding enterprise will be largely determined by how many lambs survive to weaning. Lamb mortality rates in the first week of life vary in Australia from 10 per cent in very favourable conditions to as high as 40 per cent in poor seasonal conditions where predators are a problem.

Paddocks

Lambing paddocks should have good levels of high quality pasture. Water supplies should be close to shade if lambing occurs during hot conditions. Suitable shelter should be available for lambs in cold areas. The best type of shelter will protect the new lamb from wind chill at ground level and this includes areas of tussock, clumpy grasses or low scrub to ground level. If ewes are going to be inspected during lambing, the paddock should be easy to look around with minimal disturbance to the lambing flock.

Lambing inspections

It will depend on the producer's own approach to stock husbandry whether lambing ewes are inspected daily or left to lamb unattended. If lambing inspections are done, it is important to:

- Get ewes used to a vehicle in the paddock before lambing begins.
- Never take dogs.
- Keep disturbances to an absolute minimum to avoid interrupting the ewe/lamb bond.

Recognising problems

A ewe will lamb within half to two hours after first showing signs of labour.

The main causes of difficult births are large lambs in maiden ewes, and malpresentation of the lamb (Figure 4.4). A ewe experiencing a difficult and protracted birth will show repeated straining as the uterus contracts for no result. Often the vulva area and the first of the birth membranes will be excessively discoloured by dirt.

If a decision has been made to assist ewes with difficult births, this must be done carefully. Large lambs in a normal delivery position can usually be delivered by restraining the ewe and applying gentle but firm pressure to the front legs of the lamb as the uterus contracts.

Once the lamb is delivered, place it in front of the ewe. With First Cross and British breed animals, it is possible to back away from the ewe as she rises and demonstrates normal maternal behaviour by licking and nuzzling the lamb. In most cases, Merinos will require a short period of confinement (12 hours) with the lamb before being returned to the flock, especially if the ewe has had to be chased. Small portable yards or a pen in the corner of the paddock can be used for this purpose.

Where there is a malpresentation, assess if the ewe is able to deliver the lamb in the present position with the assistance of firm pressure. If this is not possible, attempt to adjust the lamb into a better position for delivery. If the lamb can not be delivered, call for veterinary assistance or humanely destroy the ewe.

Lamb losses

Lamb losses on a property can vary from 5.0 per cent to as high as 75 per cent. Even in well managed flocks losses of 10 to 20 per cent are normal. The majority of lamb deaths occur within three days of birth and most are associated with a mismothering, starvation, and exposure complex. Other causes of lamb loss are injury and predators.

Mismothering
Any disturbances to the flock like drought feeding, lambing inspections or disturbance by large predators will increase the number of mismothered lambs. These losses will be greater in Merino flocks compared to First Cross and British breeds.

Starvation
Lambs below normal birth weight or lambs suffering some birth injury are weaker. These lambs' sucking stimulus is reduced and they are not able to follow the ewes as they graze. Poor nutrition can delay the onset of lactation in the ewe or reduce the ewe's maternal behaviour to the extent where she will not allow the lamb to suckle. Lambs born as part of a twin set are more likely to be lost from desertion, or lamb stealing by a ewe in the early stages of labour.

Exposure
A lamb has sufficient energy reserves for a short period after lambing, but if it does not suckle within the first few hours, survival rates will fall dramatically. Exposure losses are obviously higher as wind chill increases with wet, windy and cold conditions. Research has shown that using low, tussocky shelter belts significantly improves lamb survival rates under these conditions.

Injury
Lambs suffer physical damage or brain damage due to lack of oxygen during difficult or protracted lambing.

Predators
Losses from predators vary between areas and seasons. The main predators of lambs are feral pigs, foxes, domestic dogs and crows. (In many cases,

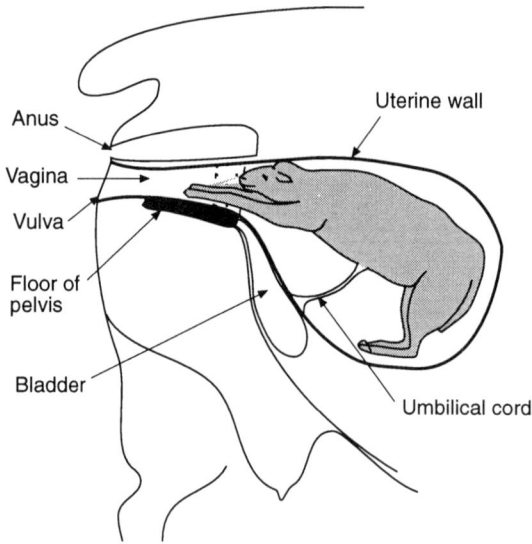

Figure 4.3 *Normal presentation of the lamb*

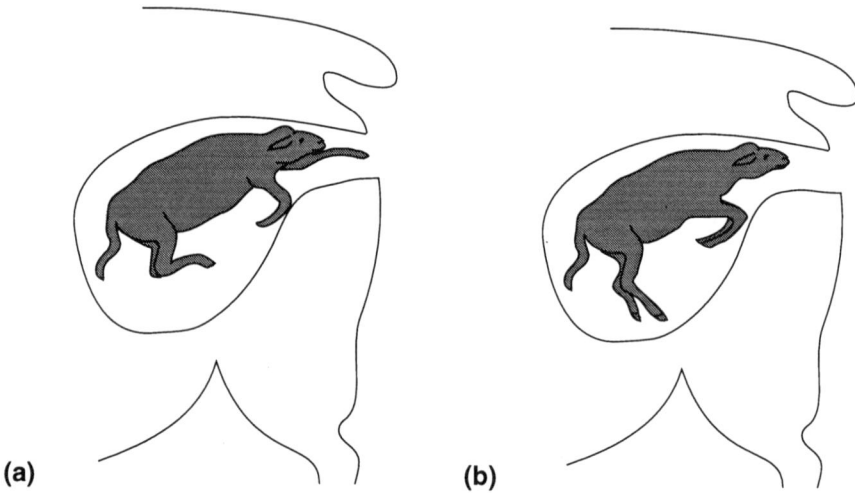

(a)

(b)

Figure 4.4 *Examples of malpresentation of the lamb*
(a) *One front leg back*
(b) *Two front legs back*
(c) *Head back*

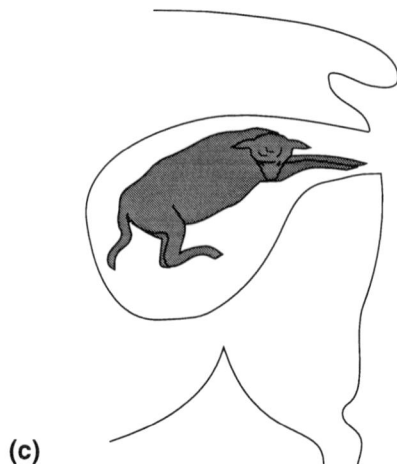

(c)

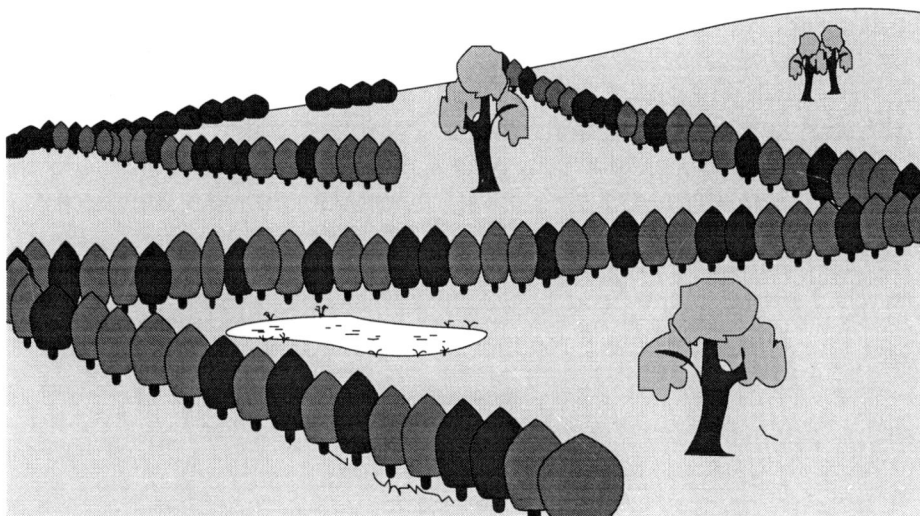

Figure 4.5 *Shelter belts*

lamb deaths are not due to attack, but predators will be seen scavenging on already dead lambs. It is impossible to tell in many cases if the predator was responsible.) In some areas heavy losses are due to the predation of feral pigs and foxes, and a controlled baiting program of several weeks before lambing and into the lambing period will help preserve lamb survival rates. Lambing at a different time to other producers in the district will increase losses from predators as the food source is limited to one area.

Determining wet and dry ewes

The identification of ewes that are not rearing a lamb is a useful management tool. The ewes are placed in a race and the udder palpated. Dry ewes are identified by the lack of udder development. Dry ewes can be taken out of the mob and feeding levels reduced.

Weaner management

Lambs are weaned from four to five months of age in most areas. This will depend on pasture conditions and whether the lamb is destined for slaughter, sale to a restocker or for inclusion in the breeding flock.

After weaning, good growth rates will be achieved under several conditions:

- Good quality pasture is available with a protein content of at least 13 per cent. Growth rates are better where there is a significant legume content in the pasture.

- Weaning paddocks have shade, water and be relatively free of grass seeds. Weaners run in paddocks with Barley Grass, Corkscrew or Spear Grass will lose condition due to seed irritation and will only improve when the source is removed by shearing.
- Animal health conditions are met. Weaners should be drenched for internal parasites and run on a worm-free pasture, vaccinated for clostridial diseases if not done earlier, and protected against external parasites especially fly strike.

Weaning is a stressful period for the lamb. Make sure the freshly weaned mob has found water. Later management and mustering can be facilitated by adding a few adult sheep (wethers or dry ewes) to the weaner mob.

CHAPTER 5

Nutrition

Ruminant digestion

The sheep is a ruminant. Ruminants are animals that have a modified gut that allow them to digest cellulose (plant fibre) that many animals are not able to break down. Ruminants graze quickly and do not initially chew their food finely. This material goes into the rumen where it mixes with previously eaten food and liquid and is fermented by microbes.

Larger food particles are later regurgitated and re-chewed (chewing the cud), re-swallowed and then passed onto the remainder of the gut for digestion and absorption of nutrients.

Ruminants receive energy from the by-products of fermentation in the rumen, protein and energy from digestion of food in the gut and by digesting the microbes that ferment material in the rumen. Protein in the diet can be degraded (digested) by microbes in the rumen or pass through the rumen (by-pass protein) directly to the sheep.

Supplementary or drought feeding of sheep must be aimed at stimulating good populations of microbes in the rumen for efficient digestion of low protein, low energy roughage as well as providing a source of energy and protein to the sheep. Any changes of feed must be done slowly (over seven to ten days) to allow the population of rumen microbes to adapt.

Nutritional requirements

Energy

A sheep's requirement for energy will vary depending on:

- Liveweight, body condition and growth rate. Heavier sheep require more energy for maintenance, while fast growth rates require higher levels of energy to be converted into body tissue.

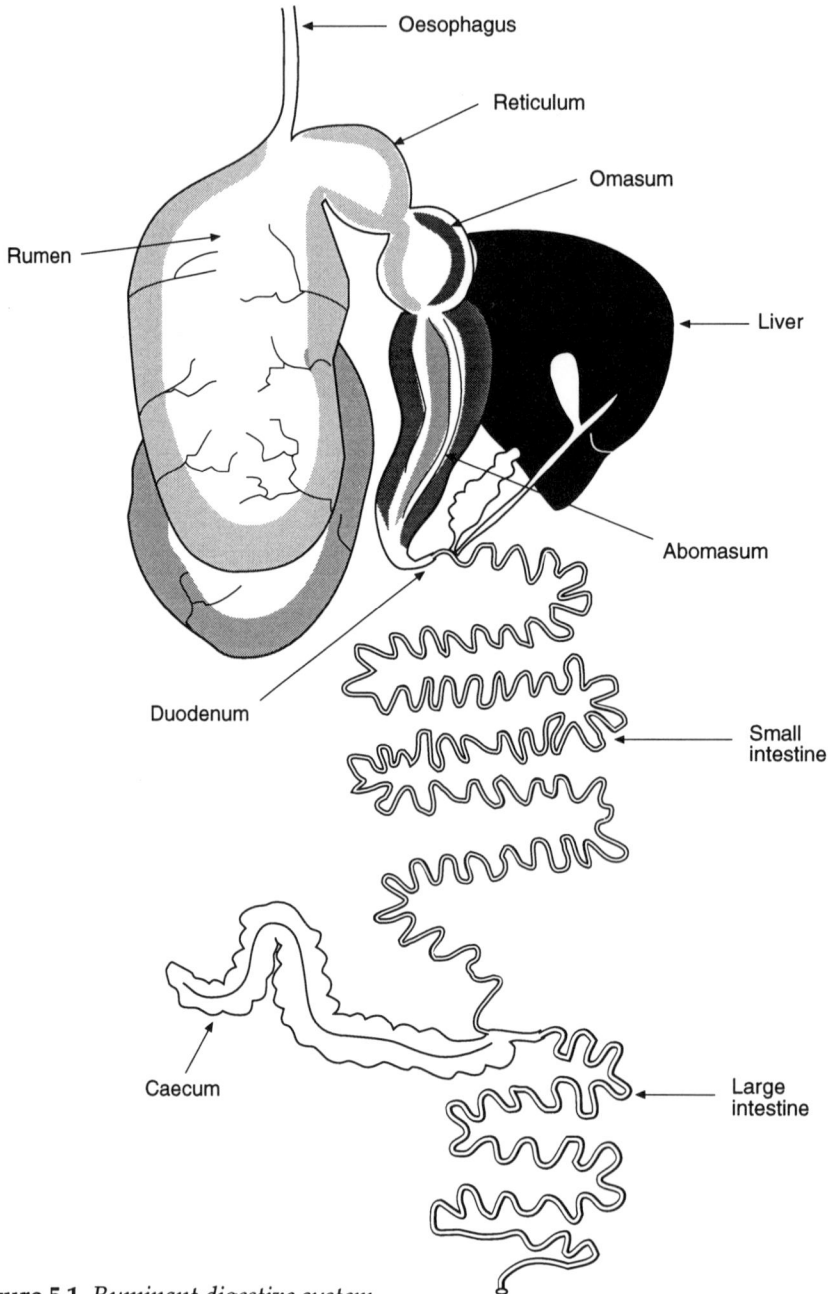

Figure 5.1 *Ruminant digestive system*

- Whether pregnant or dry, the stage of pregnancy or lactating. The energy requirement of the breeding ewe increases during the last six weeks of pregnancy, and is highest when lactating.
- Wool growth and weather conditions. Sheep have a higher requirement for energy to maintain body temperature after shearing. The amount of extra energy required will be dependent on the weather. Wet, cold and windy conditions will have a chill factor much more severe than the air temperature alone and sheep will require additional feed to survive.
- Distanced walked each day to graze will affect energy requirements.
- The digestibility of feed. Feed with a low digestibility requires more energy for digestion.

Energy requirement tables can be useful in planning energy requirements for different classes of sheep. These tables indicate the amount of energy, measured in megajoules (MJ), needed per day (Table 5.1).

Another way of comparing the different feed requirements of different classes of stock is to use DSEs. (One DSE is the amount of energy needed to maintain the body weight of a 45 kilogram dry sheep.)

This energy requirement table assumes the sheep is grazing improved pasture of medium quality.

Protein needs

Protein requirements will be different for each class of stock. A portion of a sheep's protein requirement is supplied by digestion of microbes passing into the gut from the rumen.

The protein content of poor quality pastures (mature or dead dry grass) is 2 to 4 per cent, while dry sheep require about 6 per cent. However, sheep are very efficient selective grazers. Even when overall pasture quality is low, sheep are able to select small green shoots that are high in protein and this lifts the overall protein content actually eaten (Table 5.2).

Table 5.1 *Energy requirements of various sheep types (adapted from the* Sheep Production Guide — Feeding Your Sheep)

Class of sheep	Energy requirement (MJ)	DSE requirement
Dry sheep — 45 kg	8.7	1.0
Dry sheep — 60 kg	10.8	1.3
Hogget — 30 kg and gaining 50 g/day	8.1	0.9
Hogget — 30 kg and gaining 150 g/day	12.2	1.5
Ewe — 55 kg, last month pregnancy	12.7	1.5
Ewe — as above with twins	14.6	1.7
Ewe — 50 kg, lactating	23.5	2.5
Ewe — as above, with twins	32.6	3.1

Table 5.2 *Protein requirements. (Source: Agfact A3.5.4. Drought Feeding and Management of Sheep.)*

Class of sheep	Protein requirement (%)
Dry sheep	6.0
Ewe — last month of pregnancy	8.0
Ewe -— lactating	12.0
Weaner — 25 kg, gaining 50 g/day	11.0
Weaner — 25 kg, gaining 100 g/day	12.3

Crude protein requirements are much higher when sheep are pregnant, lactating or growing. In these cases, poorer quality pastures will never meet maintenance requirements of sheep. There is a considerable range of protein content in pastures and feeds, but Table 5.3 is included as a guide.

Table 5.3 *Protein content of various feeds*

Feed	Crude protein content (%)
Young grass and legume pasture	20
Grass and legume pasture - maximum leaf area	10 to 14
Grass and legume pasture - flowering	8 to 10
Most cereal grains	9 to 12
Lupins	32
Good quality lucerne hay	16 to 20
Cereal hay	6 to 7
Oaten straw	4

Vitamins and minerals

Vitamins are usually provided in sufficient quantities by grazing pastures. Vitamin A may be deficient in situations where sheep have had no access to green feed for more than six months.

In most cases mineral requirements are also met from the sheep's normal diet. However, some areas of Australia have pastures that are deficient in minerals like cobalt, iodine and selenium. In these cases, mineral supplements may be provided by slow release "bullets" placed in the rumen. In droughts when sheep are being fed cereal grains, calcium may also be deficient. This mineral can be added as ground limestone in the grain ration.

Pastures and grazing

Plants contain large percentages of cellulose, mainly in the cell walls. This material can be fermented and digested to provide energy for the sheep's

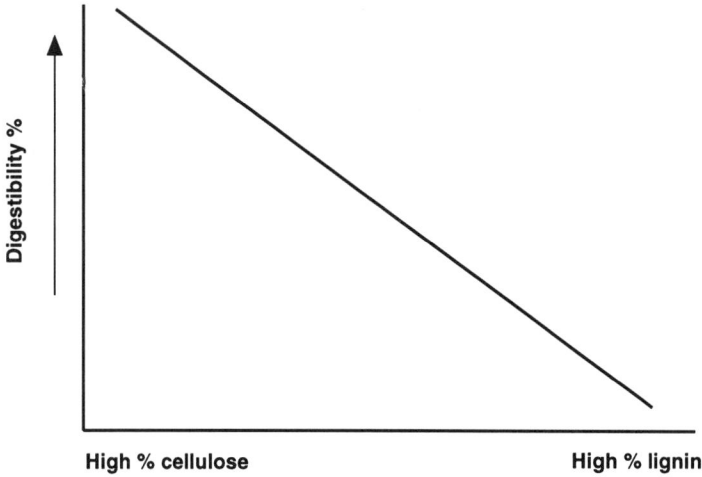

Figure 5.2 *The effect of lignin content on digestibility*

requirements. However, as the plant matures, much of this cellulose is converted to lignin and this is much less digestible.

Digestibility is a measure of how much feed value the sheep can extract from a given quantity of pasture. Obviously, the higher the digestibility of a pasture, the better will be sheep production levels.

Pastures

There are three distinct stages of growth in a pasture. Correct grazing strategies will help to maintain pasture in a productive condition.

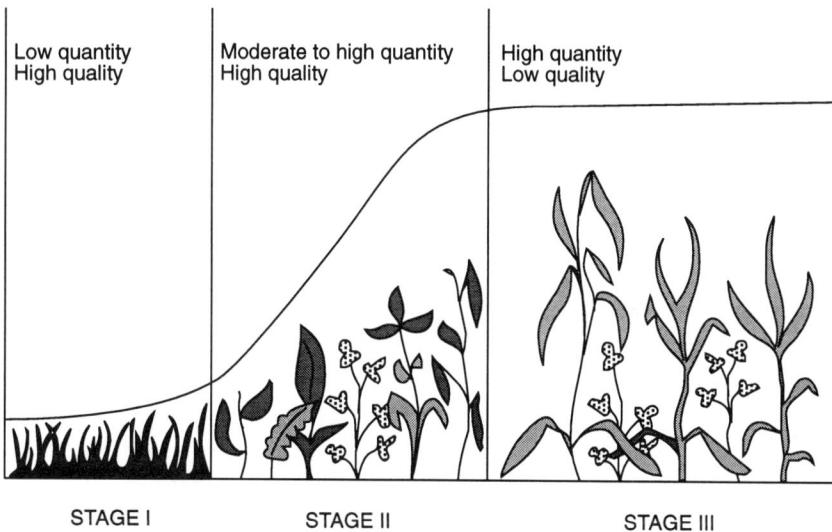

Figure 5.3 *The stages of plant growth and maturity*

In Stage I plant growth is slow because there is insufficient leaf area to trap a large proportion of sunlight. If the pasture is newly germinated, root development is shallow. The quantity of available plant material at this stage is low, and even though protein levels are very high, grazing should be avoided.

In Stage II, plant growth rates are at a maximum with all leaves trapping sunlight. Roots are well developed and extracting water and nutrients from a greater soil depth. Grazing a pasture at this stage will provide high levels of available plant material of high digestibility and good protein levels, and will provide the best mix of quantity and quality possible.

In Stage III, much plant growth has been wasted with the shading and yellowing of lower leaves in the pasture. Advancing maturity means digestibility is declining. Although available plant material is now at the highest level, this is offset by with lower protein levels and lower digestibility. This pasture should have been grazed before this stage.

Many perennial pasture species will persist better if they are allowed to flower and seed at least once a year. In this case, optimal use of the pasture is compromised for a short period to improve future production levels. Correct grazing management, where the pasture is allowed a sufficient recovery or rest period after grazing, will also improve pasture persistence and production levels.

Rotational grazing strategies

Good grazing management is aimed at grazing the pasture in Stage II. There are various rotational grazing and cell grazing techniques. The aim of these methods is to allow the pasture to grow toward the later part of Stage II and then graze the pasture quite heavily for a short period, followed by a long rest period. All pastures appear to respond well to these strategies. All plants are grazed and eaten down once. Pasture quality is also better controlled, and available plant material is more efficiently utilised, however, there are additional costs for fencing, supply of water and more time spent in stock management.

Set stocking

Set stocking involves placing a pre-determined number of sheep in a paddock for a long period, up to an entire year in pastoral areas. The reason for this is to regulate the grazing effect on the pasture.

At low stocking rates some plants will remain ungrazed for long periods and the quality of these will decline. Even at low stocking rates some plants (usually the more desirable species) may be overgrazed. High stocking rates for long periods will keep the pasture continually in Stage I, pasture production over the year will be low and desirable species like lucerne will die out of the pasture.

Supplementary feeding

Supplementary feeding is providing a feed to improve the energy or protein intake of the flock until pasture conditions improve. For example, the diet provided by pastures over summer often lacks protein. In this situation, feeding a protein supplement will increase protein intake and stimulate better fermentation of the low quality roughage by rumen microbes.

Supplementary feeds are usually provided to lambs, weaners, and ewes in late pregnancy. Methods of feeding vary but the most common methods are providing a grain supplement in troughs, laying a trail of grain or hay directly onto the ground or providing hay in feeders to reduce wastage. The principles of feeding discussed for drought feeding below can be applied to supplementary feeding.

Drought feeding

Drought feeding occurs when feeding levels are increased to a stage where the animal is receiving most of its nutritional needs from supplementary feeds, made necessary as pasture quantity declines. Drought feeding may be aimed at continuing production levels at normal levels, as may occur when finishing stock for market, but is more usually a survival ration.

Most producers cull wethers and older ewes rather than begin a feeding program with these stock, and only attempt to carry young and sound breeding stock through a drought. This allows rebuilding of stock numbers quickly when normal seasonal conditions resume. However, a realistic assessment of costs — feed requirements for each class of stock, probable length of feeding period, and expected stock deaths — should be made befor beginning any feeding program.

In a drought, the ration given to sheep will have to supply all requirements for energy, protein, minerals and vitamins. There is advice available from state Departments of Agriculture, veterinary officers and feed merchants in formulating the most cost effective rations for sheep using the various feeds available at the time.

Drought feeding methods

Introducing feeding
A drought feeding program should be started before animals fall below their survival weight. This weight will vary with different breeds. The table below is included as a general guide. Weights required will depend on whether the sheep is dry, pregnant or with lambs at foot. It may also vary with time after shearing, and expected weather conditions.

If sheep are allowed to fall below these weights, the death rate increases to unacceptable levels and significant production losses will occur.

Table 5.4 *Survival weights for sheep on drought feed. (Adapted from* Feeding your Sheep, *NSW Agriculture.)*

Type of sheep	Liveweight (kg)
Large frame Merinos, Crossbred ewes, British breeds	40
Meduim frame Merinos	35
Small frame Merinos	30

Sheep can be trained in a number of ways to accept supplementary or drought feeds. Training will usually be very easy if the mob was fed as lambs with their mothers. Mobs not previously manually fed may take a little longer to train, but training may be made easier if some previously fed sheep are added to the mob.

Begin by providing a good quality hay, spread out to allow as many of the mob access as possible. It is usually necessary to hold the mob quietly around the hay until a proportion of the mob begin feeding. When most of the mob are feeding readily, add some grain in a trail next to the hay. Increase the percentage of grain fed gradually.

Changing the amount of grain fed in the ration slowly gives the population of rumen microbes a chance to respond to the changing diet. This method will also avoid losses of sheep due to "grain poisoning". All changes in ration like increasing the amount of grain fed, the type of grain, or even changes of batches of sheep nuts must be gradual.

Figure 5.4 *Drought feeding*

Dry sheep should be fed twice weekly. Providing large feeds less often will result in most animals in the mob eating the desired amount of the ration. Pregnant or lactating ewes and young sheep should have access to feed daily.

Feeding grain, hay and nuts

Hay should be broken up to allow access to a larger number of animals. Round bales can be rolled out to achieve this result. There is a range of commercial hay racks available that will reduce wastage of feed. Several feeders are necessary for large mobs of adult sheep, to allow all animals to feed.

Grain and sheep nuts can be trailed out on the ground. Note that there will be considerable wastage of small grains like wheat if feeding is carried out in dusty conditions. Troughs made from conveyer belting, lengths of guttering or sheets of corrugated iron should be considered when feeding small grains or supplements. Larger grains like corn and processed sheep nuts have lower wastage under all conditions.

CHAPTER 6

Stock handling and sheep yards

M any operators find sheep difficult and frustrating to handle. The principles of sheep behaviour in the paddock and yards can be used to make mustering and sheep movement easier.

Sheep have a very strong innate tendency to flock together. A lone sheep placed in a paddock will travel continually in an attempt to find other sheep to associate with, and it will exhibit high anxiety levels until it does so. A sheep placed in a yard away from the mob will attempt to jump or break through back to the mob.

The "flight zone", or the area the handler must move into to force sheep to move, will vary. The flight zone of a sheep is greater when it is in a paddock by itself, compared to the same sheep in a mob where it feels more secure. Likewise, there is a narrow "point of balance", just behind the shoulder as the sheep looks forward. A sheep will tend to move forward while ever a handler or dog is behind this point, but will attempt to break to the side once this point is passed. A small move by the handler past this point will make the sheep change direction (Figure 6.1).

Mustering

A mob of sheep will show a point of balance in a similar manner to an individual. It is possible to control the wings of the mob but still keep the mob moving in a forward direction. Just how far it is possible to work a dog or move a handler alongside the mob without causing the mob to stop and circle will vary with different sheep and even the same sheep at different times. The handler needs to adapt the method to each situation.

Mustering sheep from paddock to paddock or paddock to yard varies greatly between the different production zones. In the high rainfall area,

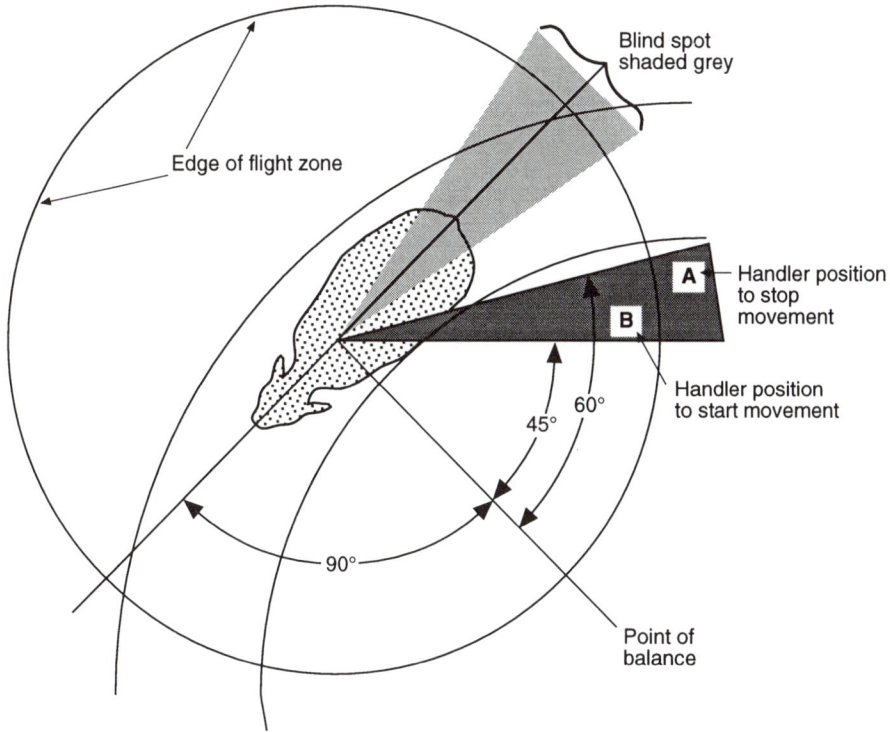

Figure 6.1 *The flight zone and the point of balance of a sheep*

the stock may be in a cleared paddock of 10 hectares and only 400 metres from the yards, but in the pastoral zone it may be spread over an area of 10 000 hectares of scrub covered land and many kilometres from the yards. Obviously different techniques will be required to muster in each of these areas but there are several principles of sheep behaviour that apply equally to each situation:

1. Sheep follow one another. This is a basic instinct that should be used to advantage, whether sheep are being mustered, moved in the yards or drafted.
2. Sheep graze into the wind. Mustering larger areas is much easier if the mob is moved into the prevailing wind. The sheep will move faster and continue to move in the desired direction for longer once the movement stimulus is reduced.
3. Sheep camp on high ground in compact mobs.
4. Sheep graze very early in the morning or late in the afternoon. Mustering large areas or difficult terrain may be easier with an early start, before the sheep graze a long way from a camp area. Mustering sheep during the middle of the day, in cool conditions when they are not camped near water, will mean finding several small mobs scattered over a large area.

5. Sheep tend to camp on easterly facing slopes, especially in cold conditions when they will often be found on these slopes until mid-morning.
6. Sheep will usually be found in compact mobs close to water in hot and dry conditions. A technique used on some pastoral properties is to close off access to all but one water point in a paddock. Over a period, all sheep in the paddock will water at the one point where they are either mustered on a regular basis or trapped around the water using a fenced funnel.
7. Sheep prefer to move up a slope rather than down, and will be difficult to move through a gate or narrow opening if it is down a steep slope.

Handling ewes with lambs

Mustering and moving ewes with lambs in a paddock must be done at a slow enough speed to prevent separation of ewes and lambs. Attempting to move the mob too fast results in most of the lambs ending up at the rear of the mob, and attempting to return to the original area to find their mothers. It may be necessary to station a person in front of the mob to slow down lead ewes and allow all lambs to keep in close and constant contact with the mob.

Ewes with lambs are always difficult to move around yards because of the tendency of ewes to continually come to the rear of the mob to find their lambs.

Extra care must be taken when stock of vastly different sizes are confined in small areas to avoid suffocation. Most husbandry operations are based on body weight, so the mob will need to be drafted. A pen-full of young lambs separated from the ewe portion then requires extra care to avoid deaths in the yard. They should be returned to the mob as quickly as possible.

Catching and holding

Catching is easier in a yard where the operator is able to move more freely than the sheep. This can be achieved by moving the mob into a smaller yard, or by forcing the sheep into a corner of a large yard, along a fence line or the corner of a paddock.

The sheep can be held standing or sitting. Sheep can be easily held standing by using the knees to force the sheep against a rail and holding the head under the chin.

To sit a sheep, approach quickly from behind, grab the sheep under the chin and place downward pressure on the rump with the other hand. Do

Figure 6.2 *Sitting a sheep*

not grab by the wool — this causes bruising which can be detrimental if the animal is destined for slaughter. An alternative method is to grab the sheep by the rear leg. When this is done, transfer the grip to under the chin as quickly as possible and sit the animal down as previously described. This will make the animal easier to control and it will be easier on your back and arm.

Sheep held in the sitting position will be easier to control if the head is kept up and the shoulder prevented from touching the ground.

Rams are usually held standing for inspections by holding both horns. The danger of the ram butting the holder will be less if the animal has been previously handled, dogs are kept away, and he is held close and is unable to rear up.

Figure 6.3 *Holding a standing sheep against a fence*

Figure 6.4 *Holding a standing sheep in the open*

Figure 6.5 *Holding a ram in the judging position. The feet should be together, the head up and the animal facing forward*

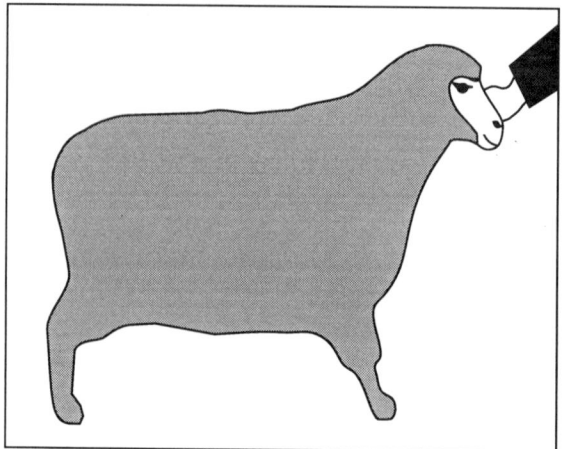

Yard work

There are several principles of sheep behaviour that should be considered when working sheep through a set of yards. Often the design of the yards will not allow all of these to be put into practice but minor changes or reorganisations can be made to reduce effort and improve work rates.

If sheep do not move well through an area of the yards, get down to the sheep's eye level. Often problems like a perceived dead end in the yards, shadows, light shafts or operator visibility become obvious.

General movement

Sheep move fastest as a group through wide and straight pens and races with enclosed sides and a clear escape point.

Fill in any side panels to prevent the sheep seeing the operator, dogs or stationary sheep in an adjacent pen. Sheep resist moving toward a visible operator and will baulk if the operator or dogs are visible in the direction they are being moved.

The sheep's instinct to follow encourages movement toward other sheep. Sheep will move through a draft or handling race better if they can see sheep ahead or treated animals escaping.

Efficient sheep handling only occurs when sheep have a clear passage to move along and a clear view of the exit with no dead ends. Provide a clear exit point for the mob, preferably toward other sheep. Sheep will stop about 3 to 5 metres from a yard set up.

Changing directions

When the direction of movement needs to be changed, sheep move best around corners in narrow races, especially if they see the animals in front rounding the corner and disappearing.

A temporary fence panel may be added to yards or forcing race to improve sheep flow. (Any curved fence panels added to the yards should be trialed; small changes can make large differences.)

Lighting

Sheep do not like moving into dark shadows, and move best from dark into light areas. Carefully plan the distribution of trees to avoid excessive shadows across the main race area.

Habitual routes

Sheep learn a route through the yards, and follow the same direction of movement.

When working sheep in conventionally designed yards, the direction of sheep flow should remain the same. In modern circular and "bugle" sheep yards usually only one direction of movement is possible.

Sheep learn quickly

Sheep that have had an unpleasant experience will be more difficult to handle the next time in the yards. Perform necessary operations with a minimum of stress to the animal, keep dogs under control or muzzled and handle sheep humanely.

Sheep will flow better if they are not moving directly toward the shearing shed, handling race or another site of a previous stressful experience. Animal flow is best in these situations where the direction of flow is turned quite sharply at the last moment into the shed, handling race or the sheep handler.

Dogs

Good sheep dogs used in yards reduce effort and stress levels of the handler and improve work efficiency. Poorly trained dogs or dogs not under control in the yards definitely increase the work effort required and raise stress levels in handler and animal. Biting dogs can cause extra work treating wounded animals against flystrike for the following weeks.

Forcing

Sheep should be moved around the yard with sufficient firm pressure to keep the mob moving. Force (stimulus to encourage movement) is more effective when applied to sheep toward the lead of the mob. When lead animals have started moving, the remainder of the mob will follow with very little effort on the part of the handler. Continually applying excessive force is tiring for handler and dogs, and will cause some animals in the mob to fall over or jam in bottlenecks and gateways, and reduce the flow of the remainder of the sheep. Sheep that go down may be bruised and may smother.

Pressure used in forcing areas of the yards, like the entry to the handling race and the draft race, should be firm. Start a smaller portion of the mob moving where the same amount of force will have a far greater impact, and then use the following behaviour to encourage the remainder of the mob. With difficult sheep, the operator is often best working at the front of the mob to add force while a well controlled dog keeps the remainder of the sheep in a position to see other sheep ahead moving past and away from the operator.

Figure 6.6 *A race holds sheep securely during husbandry operations*

Filling a race

Start sheep moving along the race to the very end where possible. Sheep visible at the end of the race will assist with filling. When the race is nearly full of loosely packed sheep, walk back along the race or use a well trained dog to tighten the sheep. This increases the number of sheep in the race and makes husbandry jobs easier as the sheep's movement is restricted.

Check the density of sheep in the race so that none go down and smother, or miss being treated. A gate should be used to divide long races after filling to minimise the chances of losing sheep.

Drafting

Most yards have a facility to draft, or separate, sheep two or three ways. Efficient drafting rates require a steady rate of sheep moving through the draft at a speed the operator can handle. Sheep will start to flow through the draft more easily if the drafter steps back a little and the gate is operated smoothly and quietly.

The speed of flow can be adjusted by quite subtle movements of the drafter to increase or reduce visibility. A movement in or out from the end of the draft has a great effect on speed of sheep flow.

The sheep's following instinct can be used to make sheep flow more continuous, by drafting the larger portion of the mob straight ahead.

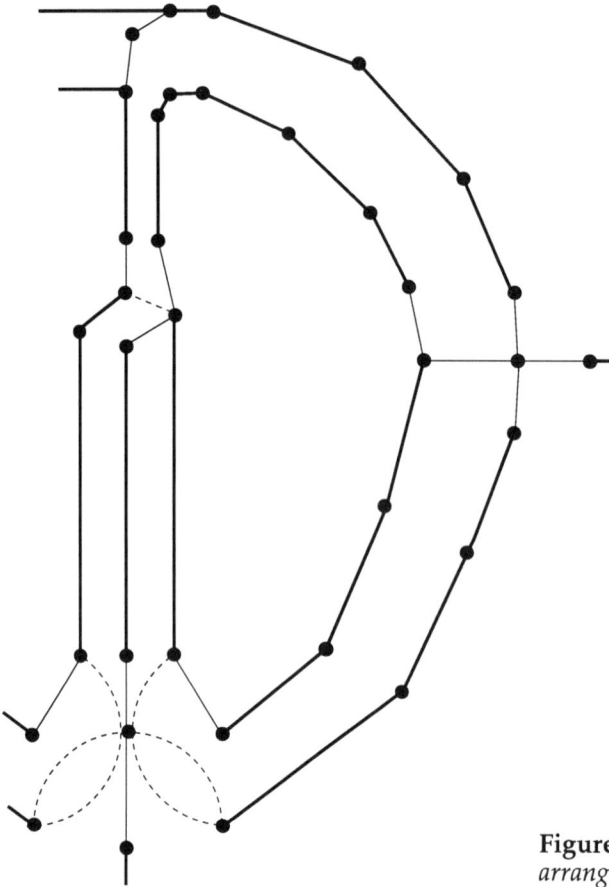

Figure 6.7 *A three way drafting arrangement*

Using sheep handlers

Using a sheep handler takes time to learn the best and most efficient method of getting good work rates. Here a few general recommendations:

1. Do not be afraid to experiment with different set-ups, forcing yard angles and position of the handler to find a combination that works in your yards.
2. Complete husbandry operations as quickly as possible while the animal is in the handler.
3. Do not leave individual animals in the handler for longer than necessary.
4. Do not use dogs that bite, excessive force or electric prodders anywhere in the yards, and especially not to feed the handler. Sheep remember unpleasant experiences.
5. Allow treated stock to return to a non-threatening situation (back with the mob in a large yard) as quickly as possible.
6. Sheep will often work better through yards if they have been locked

up the afternoon before. Remember, never lock up heavily pregnant ewes and all stock should be allowed access to feed and water as soon as practicable after husbandry operations have been completed.

7. Sheep work best through all yards, including sheep handlers, in the early morning or late afternoon. Avoid working sheep during the hottest part of the day.

8. If the situation is causing you stress, do not knock the sheep around — they will remember and be harder next time. Do not knock your dog around — it will go home and you will do the job by yourself. Do not knock yourself around — knock off and have a rest!

Yard design

Sheep yards and shearing sheds are the most important structures on any sheep property. Well designed and constructed yards make sheep work a very labour efficient operation where high work rates can be achieved. Good yards reduce the stress of handling on both the sheep and the operators.

General factors

Location

Carefully consider the following when selecting a location to build a new set of sheep yards. It is difficult to satisfy all of these factors in selecting a site but the most important in each situation can be applied.

- Yards should be in a central location on the property.
- Location should work in with existing sheep handling structures, such as the shearing shed or sheep dip.
- Access to power may be required.
- Water supply should be available at the site.
- Slope. Steep slopes can make sheep movement difficult while flat sites have poor drainage. A gentle slope is preferable.
- Use of existing trees for shade.
- Surface. A good hard surface is essential as sheep's hooves are very abrasive. Gravel may need to be carted to the site to improve the surface.

Size

This includes the size of both the sheep working area and holding areas. One factor that may limit size of the yards is the cost of the construction and materials. Where budgets are limited a smaller sheep working area can be used in conjunction with holding yards and paddocks, but this will necessitate the sheep being worked in smaller mobs.

Yard size is usually calculated so that the yards will accommodate the largest mob of sheep on the property. Once this mob size has been established then the following densities can be used as a guide to the size of the various parts of the sheep yards:

- Forcing yards should allow 3 sheep per square metre.
- Holding yards allow 1 sheep per square metre.
- Working race to hold 30 to 50 sheep, from 3 to 3.5 metres long.
- Forcing yard to hold one to two times the capacity of the working race.

Note: the yard design needs to be considered when calculating sizes as some designs are more efficient in working a set number of sheep.

Materials

The material used will be determined by availability, price and construction method.

Hardwood timber is the strongest and most durable timber for yard building. It is available in most areas and allows easy construction. Hardwood is highly susceptible to white ant (termite) attack and needs to be treated in areas where white ants are prevalent. One option to reduce white ant attack is to use Cypress pine posts and hardwood rails. Cypress pine offers good natural white ant resistance but is not as strong as hardwood.

Steel is the most popular material used in yard construction. It offers greater strength, and outlasts all other materials. Steel comes in a range of forms suitable to sheep yard construction: pipe, RHS, W-strap, sheet and mesh can be used in various combinations to obtain the best structure for each situation. W-strap is the best for yard rails as it is highly visible. Most steel suppliers sell second grade material in the above categories which is adequate for yard building, and offers a discount over first grade material.

Rubber from second hand conveyors rubber can be used to fill in yard panels instead of steel sheeting. It is much cheaper than steel, offers good visibility and is flexible if sheep run into it. Conveyor belt rubber is not as permanent as steel and is only recommended as a cost saving measure.

Design features

The following yard elements can be used with all types of yards.

Drafting race

The aim of the drafting race is for the operator to be able to individually identify each animal while in the race to allow it to be drafted into the desired pen. Drafting races can have either two or three drafting gates into separate holding pens, depending on individual requirements. The length of the race is critical, as a short one makes identification difficult while an overly long drafting race will slow the flow of sheep

The drafting race should be 3 to 3.5 metres long with enclosed sides and be 850 to 900 millimetres high. A "V" shaped race is preferable, as it restricts the flow of sheep down to a single file and prevents sheep "packing

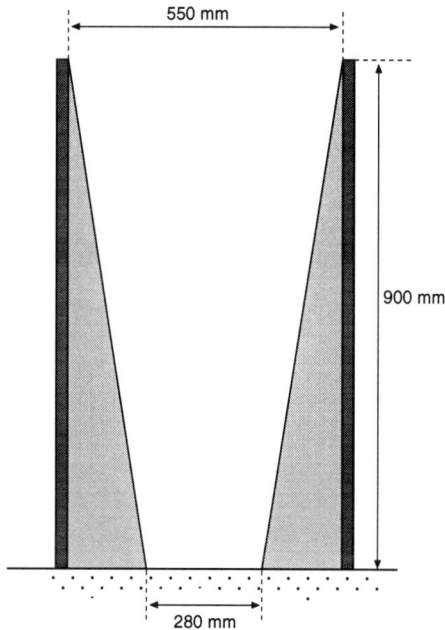

Figure 6.8 *Drafting race*

up" at the beginning of the race. The race should be 280 millimetres wide at the bottom and 550 millimetres wide at the top. Drafting gates need to be approximately 1200 millimetres long with open rails to allow visibility straight ahead.

The base of the drafting area must be cemented as this area is highly susceptible to erosion. If possible the race should be running slightly uphill and away from the shearing shed to improve sheep flow. Shadows in the drafting race cause sheep to baulk so the race should face north or south to prevent shadows from the sun.

Working race
The working race is similar to the drafting race. It is frequently used and if designed and constructed properly will allow easy and efficient sheep work.

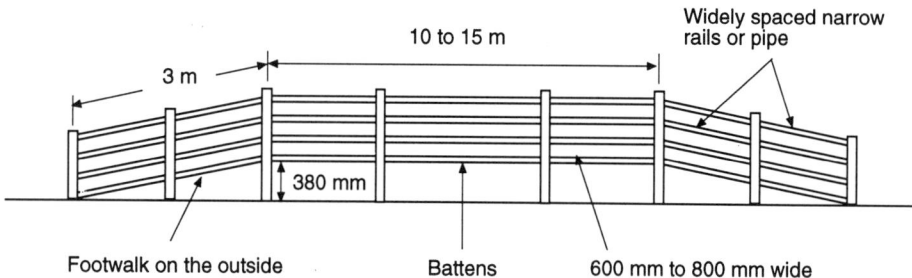

Figure 6.9 *An elevated classing race*

The design of the working race is determined by the personal preference of the operator and the type of operations to be carried out.

The working race should be 10 to 15 metres long, 600 to 800 millimetres wide and 850 to 900 millimetres high. It can be either a single or double race depending on the numbers of sheep to be processed. The theory behind a double race is that sheep are left in one side of the race to draw in the sheep entering the adjoining race. Some working races are designed with a drafting gate at the end to allow sheep to be drafted after classing or mouthing operations.

The surface under the race should be cemented to prevent erosion and allow drainage while jetting. A covered race is an advantage in adverse weather conditions offering protection to both the sheep and operator.

Forcing yards
These yards need to be very strong as they have to withstand a lot of pressure. It is recommended that the posts be 2 to 2.5 metres apart and cemented into the ground 750 to 900 millimetres deep. The size of the forcing yard should be one to two times the capacity of the working race and the sides should be filled in with steel sheeting to improve sheep flow.

Holding yards and paddocks
These hold the sheep before and after handling. Holding yards should be strongly constructed and allow enough space for 1 square metre per sheep. Holding paddocks can be built from normal fencing materials with post spacings moved closer together. They can vary in size but most are 0.25 to 0.5 hectares. (The use of additional holding paddocks and fewer holding yards is one means of reducing overall yard costs.)

Other considerations
- Fresh drinking water should be made available in holding yards and paddocks.
- Sprinkler systems are advantageous throughout the yards in hot dry climates.
- Shade trees are recommended in holding yards and paddocks.
- Wind breaks may be necessary beside the yards in some areas where high winds are prevalent.

Yard plans

There are three main types of yards that can be used to suit the individual needs. These are the bugle, circular, and rectangular yards.

Bugle yards
The bugle design is based around a circular, slowly tapering sheep forcing area. It is designed on the principle that sheep are good followers, and will

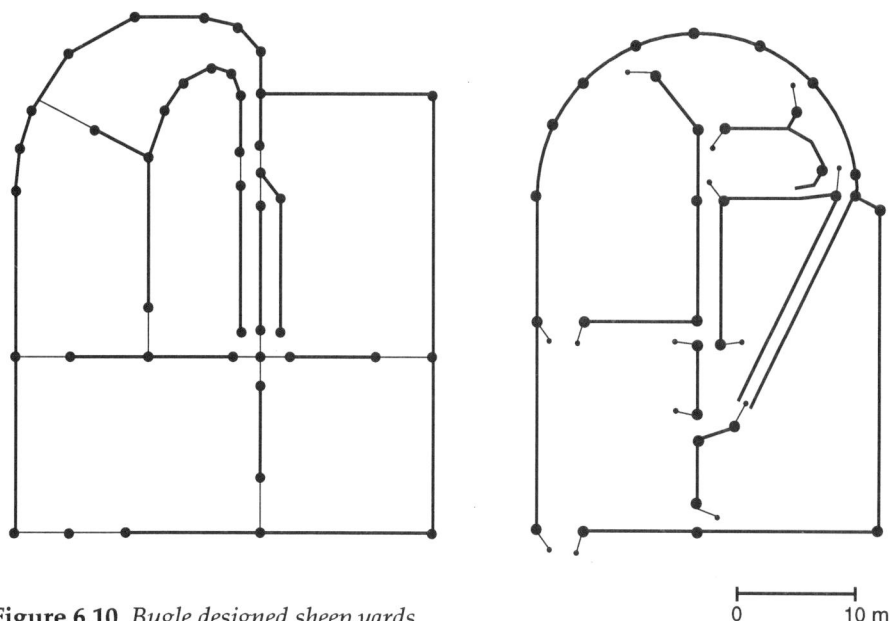

Figure 6.10 *Bugle designed sheep yards*

0   10 m

follow other sheep disappearing into the "bugle". There are many variations of this design and the bugle can be either a left or right hand turn.

The advantages of bugle design are:

- Improved sheep flow as sheep only move one way through the yards.
- Easier for the operator to move around the yards.
- Drafting can be done by one operator.
- More efficient use of space as entry yards can also be used to hold
  sheep after drafting.
- Many facilities can be connected to one bugle forcing area; such as
  the loading race, shearing shed, sheep handler and dip.

The disadvantages include difficulties to plan and construct the bugle
section and the added expense of the more sophisticated design.

Circular yards

These can be either full circle or semicircular in design. They work on the
same principles as bugle yards, and the advantages and disadvantages of
each are similar to those listed above (Figure 6.11).

Rectangular yards

This traditional yard design has been used extensively throughout the Australian sheep industry. Many older yards constructed in this design have
been poorly planned and built, leading to excessive criticism of rectangular
yards. However, most yards built today incorporate modern designs, as it

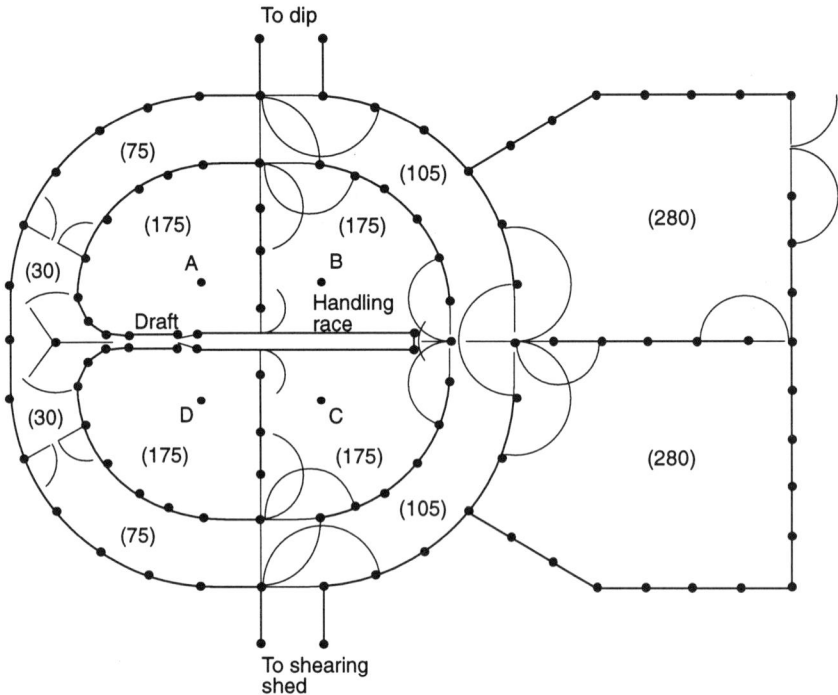

Figure 6.11 *An example of a circular yard plan*

has been proven that sheep move better in circular patterns and yards with fewer corners improve sheep flow.

The advantages of rectangular yards are that they are simple to design and build, and are good for small sets of yards, and portable yards. Disadvantages include:

- Difficult for one operator.
- Poor sheep flow compared to other designs.
- Often inefficient use of holding yards.

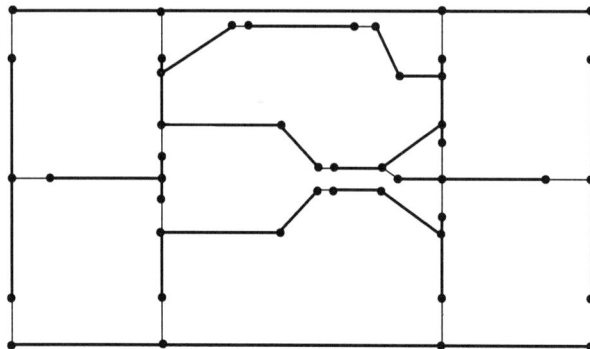

Figure 6.12 *A rectangular yard design*

CHAPTER 7

Health and diseases

External parasites

Sheep lice

Lice are a major problem and cost the Australian sheep industry many millions of dollars a year in lost production, downgrading of wool clips and control costs.

Sheep lice are small tan coloured parasites that live on the skin of sheep. Often referred to as biting lice, these small parasites cause the skin of the animal to become very itchy. This induces the sheep to rub and bite at the affected area. As a result, wool production decreases and the wool becomes damaged and matted, reducing its value.

Symptoms of lice infestation are very obvious with sheep rubbing on trees and fences and biting at their wool. The fleece appears straggly and matted. Lice can be seen by opening the wool on the affected area in bright sunlight. They will be detected on the skin as a small parasite approximately 2 millimetres long. Lice do not like direct light and will move quickly once the wool has been opened, making them difficult to detect.

Sheep lice are spread by body contact among animals. It is important to restrict stock movements once lice have been detected to prevent further spreading.

Lice can be suppressed until shearing by jetting the affected mob at regular intervals until shearing. Alternatively, there are pour-on back line treatments available that can be used to eradicate the lice. Dipping long woolled sheep is effective in eradicating lice but is usually avoided because of the downgrading of the wool clip due to dip stain.

Once shearing has been completed sheep should be thoroughly dipped or a back line treatment applied to eradicate the lice. Ensure that paddocks are cleanly mustered and fences are in good repair to eliminate further

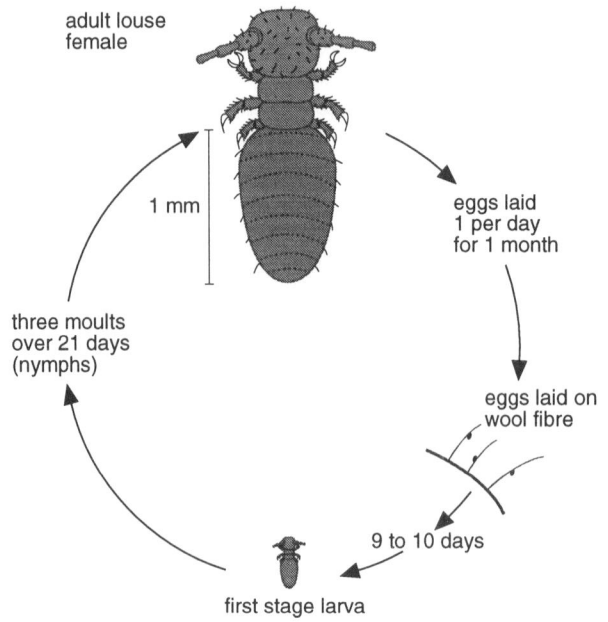

Figure 7.1 *Life cycle of sheep lice*

Figure 7.2 *Lousy sheep*

problems. All sheep on the property should be dipped after shearing if lice were detected at shearing. It is difficult to control and eradicate lice on properties running flocks of sheep shorn at different times. Always quarantine and thoroughly check any purchased sheep for lice infestations.

Itchmite

Itchmites are a microscopic parasite that live on the skin of sheep. They build up slowly over several seasons and signs are similar to that of lice infestation. Itchmite are not visible and can only be detected by examining skin scrapings under a microscope. Many of the dips (but not all) used to control lice after shearing will eradicate itchmite. Dips that eradicate itchmite need to include Roterone. Drenching with an Ivermectin based product is another effective method of eradicating the itchmite.

Keds

Sheep keds are only a minor problem in the sheep industry today, due to the effectiveness of modern chemicals.

These large blood sucking parasites are often confused with ticks. They are usually 3 to 6 millimetres long and found either on the skin or just above it in the wool. Keds cause sheep to become anaemic and generally unhealthy. Treatment is by dipping with a recommended chemical.

Blowfly strike

Blowfly strike causes major losses in the sheep industry each year. These losses are mostly production losses, but in severe cases sheep will die if not treated.

Sheep with flystrike are easy to detect as they normally stand away from the mob, with their head down and can be seen biting or rubbing at the infected area. Flystrike appears as a wet, dark and greenish stained area of the wool.

Sheep blowflies are unable to strike dry, healthy skin. The blowfly maggot requires warm moist conditions to survive with a ready supply of liquid protein for food. There are several areas of the sheep which are commonly attacked:

- Body strike starts in an existing area of skin damage, usually an area of fleece rot, dermatitis or a wound.
- Breech strike in the area around the tail and crutch, caused by soiling of the wool with faeces and urine.
- Pizzle strike occurs in rams or wethers around the pizzle and spreads to other areas of the body.
- Poll strike is most common in rams and occurs between the horns or on the top of the head due to fighting.

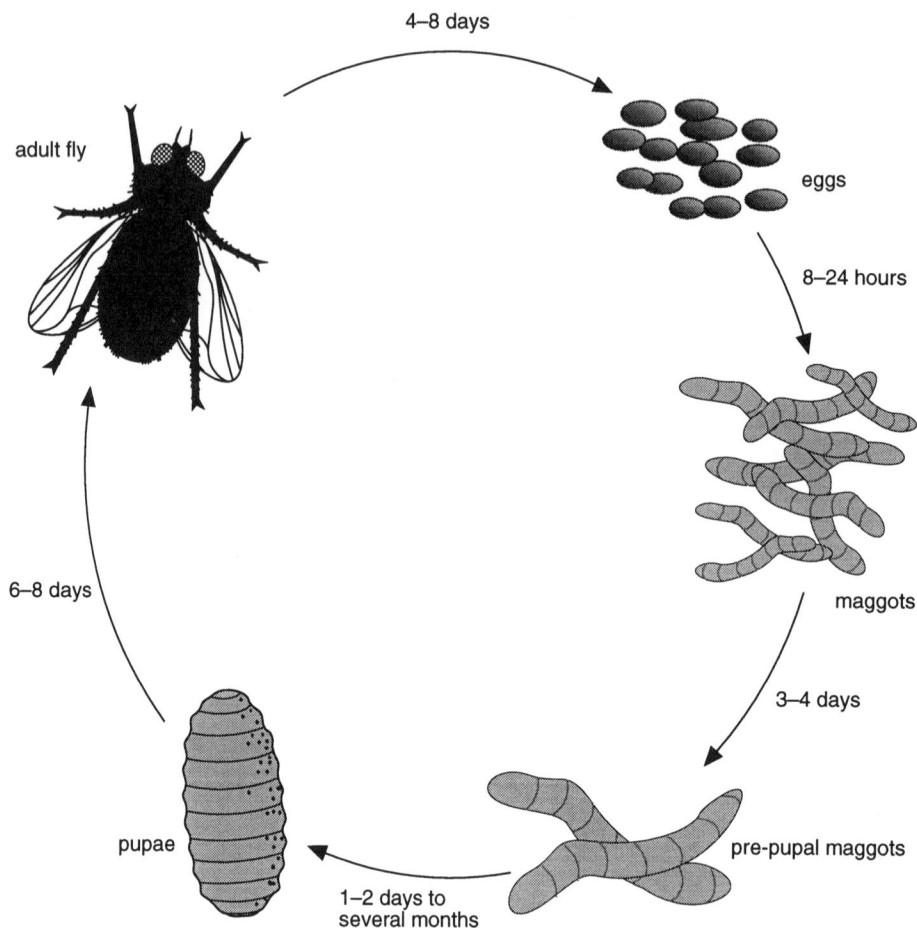

Figure 7.3 *The life cycle of sheep blowfly*

There are two types of fly strike:

1. Primary strike. The fly causing the initial strike targets a moist or stained patch of wool and lays its eggs. These eggs hatch producing maggots which attack the flesh and soon spread over a large area. Primary strike maggots are white and smooth.
2. Secondary strike. If a primary strike goes untreated for several days then secondary blowflies may attack the infected area. These are much more severe than primary flies and maggots of the secondary fly will kill sheep in only a few days. Secondary fly maggots are dark brown and hairy in appearance.

Infected sheep are usually treated with an organophosphate based chemical. If there is only a small number to be treated they can be caught individually, wool shorn from the infected area and the chemical applied. In cases where numbers are too great to treat by hand, sheep will need to be yarded and treated by jetting.

Prevention and timely and effective control of flystrike will limit the number of adult flies available to find and strike susceptible sheep. Most adult flies travel less than 2 kilometres from where they hatch. Severe "flywave" problems on a property usually result from a failure to firstly identify and control small or covert strikes and then poor preventative and control measures.

There are several options in preventing flystrike:

Jetting

Jetting susceptible sheep prior to high risk seasonal conditions is a standard management procedure in many flocks. Weaner and hogget sheep will be more susceptible to flystrike than adult sheep. Modern chemicals have made it possible to gain up to three months' protection after jetting. In most cases this will get sheep through a bad period without a second treatment. Sheep can be jetted by hand using a jetting wand or run through an automatic jetting race. Hand jetting is much slower but results in a longer period of protection.

Crutching and mulesing

Crutching prior to flystrike risk periods (usually October and February) is another management method which can be used against flystrike. Crutching will virtually eliminate the possibility of pizzle and breech strike for two to three months, and reduces poll strike in rams and wethers.

Correctly mulesing and marking lambs will reduce their risk of flystrike for life.

Shearing

Shearing can be timed so that wool length is as short as possible during the main blowfly danger period in your area. If sheep are shorn prior to the main flystrike risk season, problems will be greatly reduced. This means the sheep will have about six months' wool at the time of the next main flystrike season when they will usually be crutched.

Breeding and selection

The main cause of body strike is an existing condition of fleece rot or dermatitis. Culling or treating sheep affected with dermatitis and fleece rot, having excessive body wrinkle or obvious conformation faults in the shoulder area will reduce the incidence of body strike. Resistance to fleece rot is an inherited trait, so selecting sheep from strains or bloodlines where fleece rot has been reduced will reduce the overall susceptibility to body strike.

75

Internal parasites

Worms

Internal parasites require moist conditions to survive whilst in the larval stages and therefore create greater problems in high rainfall areas. This means that arid parts of Australia with under 350 millimetres of rainfall do not generally experience worm problems. Figure 7.5 outlines the parts of Australia where worms are a major problem in sheep flocks.

There are four main types of worms which infect sheep:

1. Small brown stomach worm.
2. Black scour worm.
3. Thin necked intestinal worm.
4. Barber's pole worm.

These types of worms all have a similar life cycle. Adult worms attach themselves to the inside of the gut. The female lays eggs which are carried out of the digestive tract in the sheep's faeces. These eggs remain on the pasture until conditions become warm and moist when they hatch out to become first stage larvae. The larvae feed on bacteria in the faeces, growing

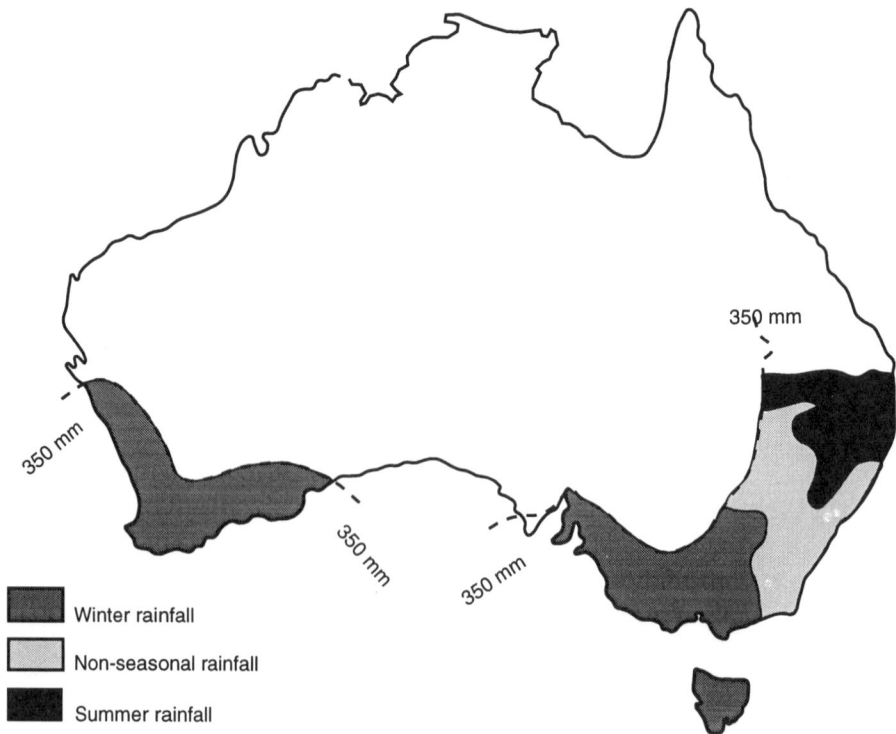

Figure 7.5 *Areas of Australia commonly affected by worms*

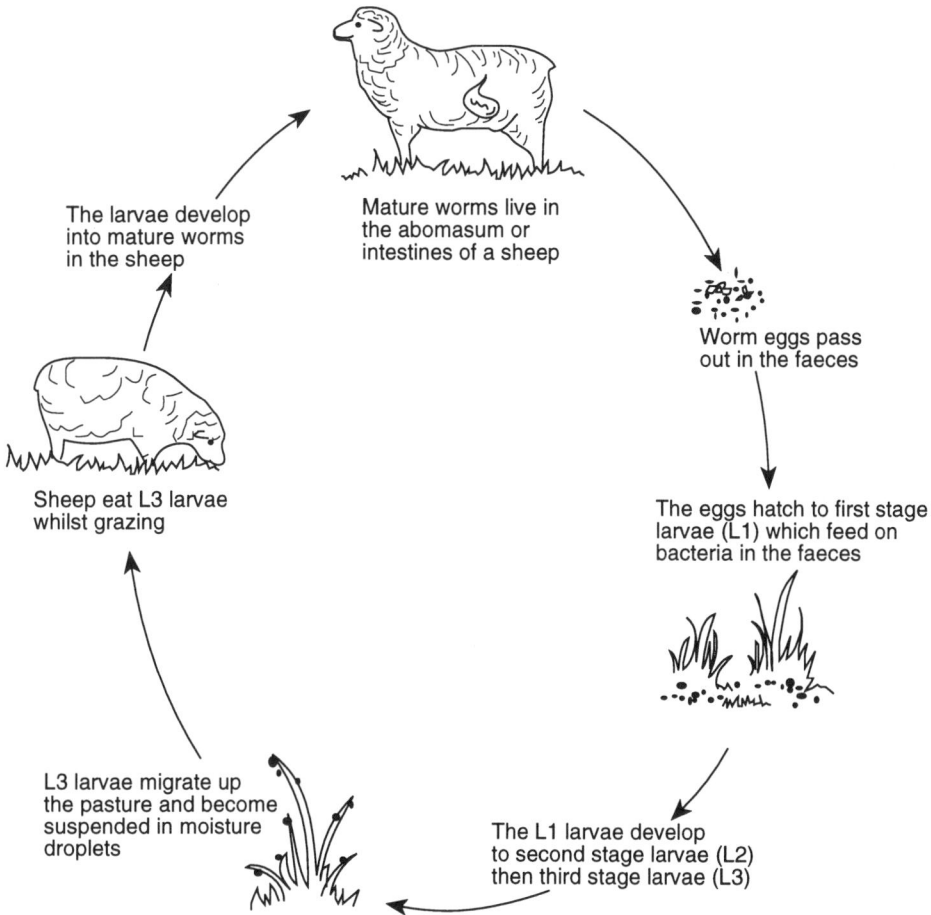

The larvae develop
into mature worms
in the sheep

Mature worms live in
the abomasum or
intestines of a sheep

Worm eggs pass
out in the faeces

Sheep eat L3 larvae
whilst grazing

The eggs hatch to first stage
larvae (L1) which feed on
bacteria in the faeces

L3 larvae migrate up
the pasture and become
suspended in moisture
droplets

The L1 larvae develop
to second stage larvae (L2)
then third stage larvae (L3)

Figure 7.6 *The life cycle of parasitic worms in sheep*

to become second and eventually third stage larvae. Once the larvae reaches the third stage they move up the leaves of the pasture, where they are eaten by grazing sheep. These larvae then grow into mature worms in the stomach of sheep and the cycle recommences.

The first signs of worms in sheep are a loss of condition and scouring. Reductions in weight gain and wool growth are also associated with worms but are hard to assess in the early stages. Sheep may have pale skin and gums due to the anaemia caused by worms. It should be stressed that if these signs of a worm problem can be seen, a significant loss of production has already occurred.

The most effective method of assessing worm populations is to do a faecal egg count. This is performed by collecting a small sample of fresh faeces and sending it to a laboratory where the number of worm eggs can be counted using a microscope. This result can be used to determine whether worm populations are high enough to require drenching.

Drenching programs need to be part of an overall management strategy to ensure maximum results in worm control. Integrated management/ drenching programs have been developed by the various state agriculture departments. Two programs developed in New South Wales are Wormkill for summer dominant rainfall areas, and Drenchplan for winter dominant rainfall areas.

Integrated programs have several common elements:

- A dose of drench just before summer to reduce larva levels on pastures.
- Using hot and drier conditions over summer to further reduce larva.
- Another dose of drench at the end of summer to clean out any worms picked up over the summer period.
- Providing pastures with low worm populations for most susceptible stock (lambs and weaners). Suitable pastures would be those used for cropping, cut for hay, or only grazed by adult sheep and cattle since the previous drenching.
- Young sheep should not be grazed on lambing paddocks until conditions have reduced worm larva contamination.

There are many drenches commercially available. These can be said to be either broad or narrow spectrum drenches, depending on the number of targeted parasites. The broad spectrum drenches can be further divided into three groups:

1. Benzimidazole group (white drenches); for instance Panacur, Valbazen, Systamex, Closal or Rycoben.
2. Levamisole group(clear drenches); for instance Nilverm, Ripercol, Nilzam and Citamin.
3. Ivermectin group including Ivomec, Cydectin and Oramec.

Narrow spectrum drenches include Razar and Seponver.

Worms in a particular area can become resistant to one group of drenches. This requires changing or rotating drench groups to prevent resistance. It is recommended to carry out a drench resistance test regularly. This test measures the amount of resistance worms have to the drench being used, and can be used to determine whether the current drench is effective or needs changing.

Liverfluke

Liverfluke is an internal parasite approximately 25 millimetres long shaped like a leaf. It lives in the ducts of the liver, hence its name. Liverfluke is limited to high rainfall areas where a small freshwater snail occurs, as this snail is part of the life cycle of the fluke.

Liverfluke affect sheep of all ages. The most obvious signs are very sick or dead sheep, but in less severe cases sheep lose appetite, develop anaemia and are generally weak and lethargic. Diagnosis is by examination of the sheep's faeces under a microscope to look for eggs or by post-mortem examination of the liver and intestines to inspect for fluke.

The best time to drench for liverfluke is during autumn and early winter. This minimises pasture contamination with fluke eggs. A second drench may be necessary later in the year but this needs to be determined by further faeces testing. It is important to use a drench that kills both immature and mature fluke. Fasinex is a good example of a modern drench that achieves this result.

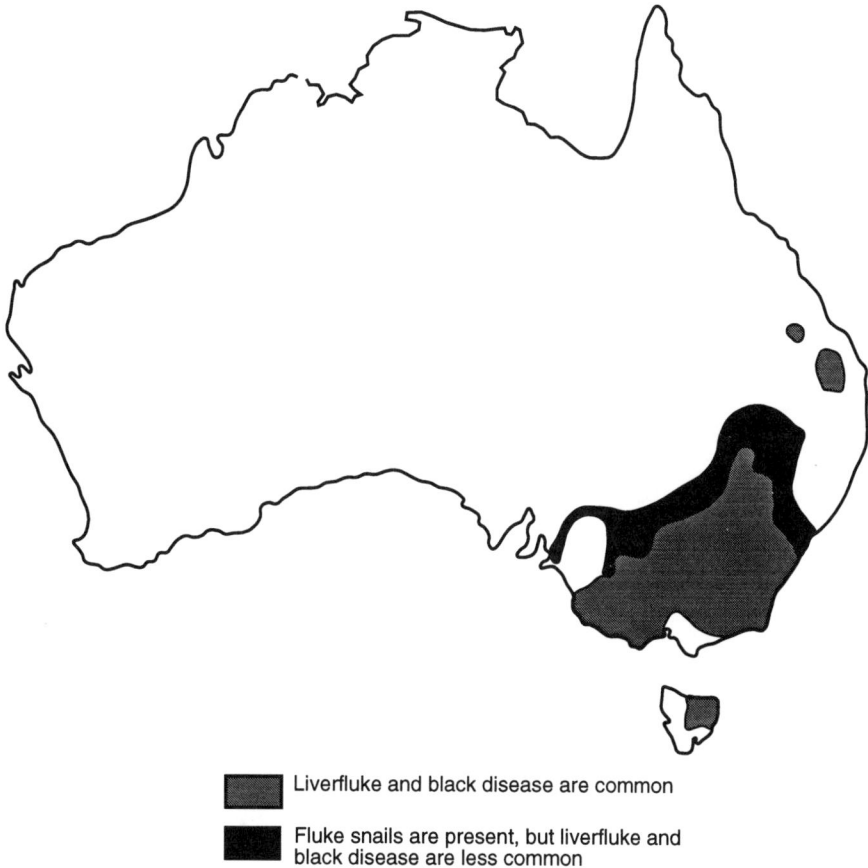

Liverfluke and black disease are common

Fluke snails are present, but liverfluke and black disease are less common

Figure 7.7 *Australian liverfluke areas*

Mature liver fluke produce thousands of liver fluke eggs. These pass out in the faeces

AUTUMN - WINTER

The metacercariae break down within the intestines, and the immature fluke which emerge migrate to the bile ducts in the liver. Development to maturity takes 10 weeks

If temperatures are warm and there is moisture on the ground, the eggs hatch to produce miracidia

The miracidia swim until they find and burrow into a *Lymnea tomentosa* snail

Sheep and cattle become infected by grazing pasture which contains metacercariae

SUMMER - SPRING

Within a snail, each miracidium develops into hundreds of cercariae

The cercariae leave the snail, attach to the grass blades, and form cysts around themselves. The cysts are knowm as metacercariae

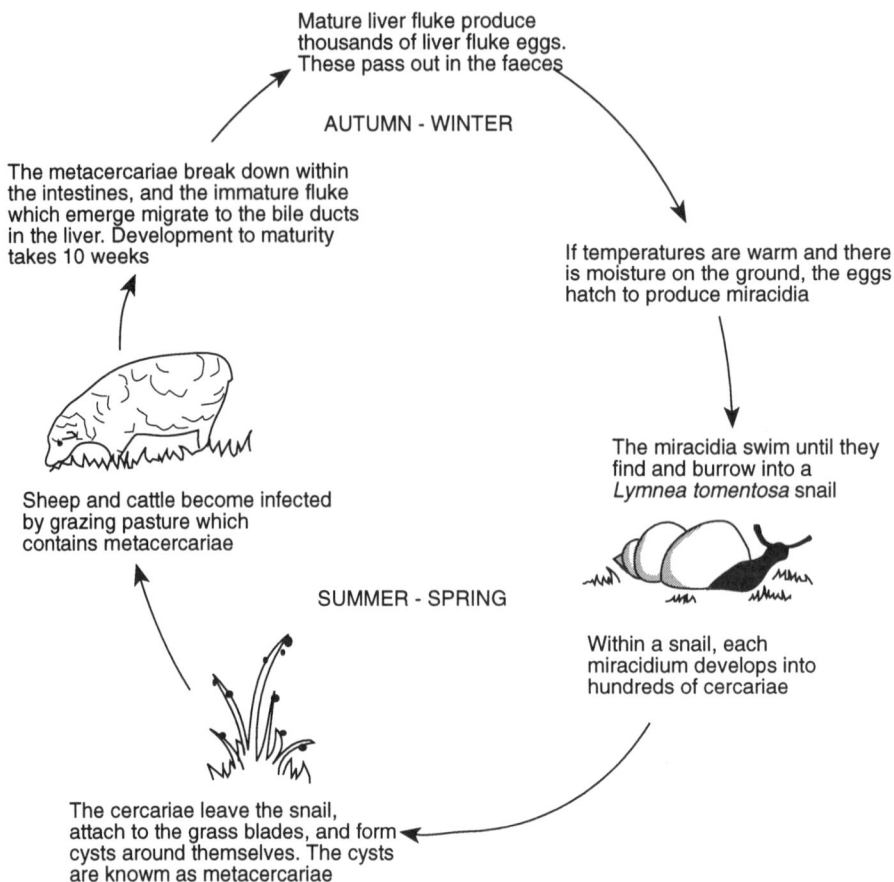

Figure 7.8 *Liverfluke life cycle*

Clostridial diseases

These diseases are caused by the clostridial family of bacteria. The diseases which cause most problems in sheep are:

- Blackleg
- Pulpy kidney (enterotoxaemia)
- Black disease
- Tetanus
- Malignant oedema.

Clostridial bacteria naturally occur in the soil and can survive for long periods without entering the body of animals. These bacteria grow where there is little oxygen and when in the body of animals they produce deadly toxins which kill the infected animal quickly.

Treatment: due to the severity of the toxins there is very little that can be done even if symptoms are noticed prior to death. The best treatment is prevention by a well planned and implemented vaccination program.

Prevention: good management practices are the best way to prevent outbreaks of clostridial diseases. Ensure that marking and mulesing equipment is clean, and use temporary yards at lamb marking time if possible. A vaccination program using a "5 in 1" vaccine (covering the five diseases) will protect the flock by developing immunity.

Vaccination and immunity

When lambs are born, they are provided with a range of antibodies in the colostrum (first milk) from the ewe. This "passive" immunity lasts for several weeks, but the lamb will become susceptible to disease as the effectiveness of these antibodies wears off. The level of protection from the ewe to the lamb can be increased by ensuring the ewe is fully vaccinated and has a high level of immunity herself. A pre-lambing booster vaccination is given to ewes for this reason.

Animals develop an "active" immunity after they have been exposed to a disease and develop antibodies to the specific organism. Vaccination works by stimulating the animals' natural defences to produce antibodies. When the animal is exposed to the disease at some time in the future, antibodies can be produced very quickly to overcome the infection.

Vaccines can be "live" vaccines (like that for scabby mouth), but are more often treated by "dead" vaccines, as is the case for vaccines against clostridial diseases.

The first vaccination, usually at lamb marking, gives a low level of immunity for a short period and falls below a protective level very quickly.

The second vaccination dose four to six weeks after lamb marking gives the lamb a much higher level of immunity. This level decreases very slowly over the next 12 months.

A booster dose given after 12 months restores immunity to above the protective level for a long period. Annual boosters for all adult sheep will keep immunity levels high and give good protection to lambs from birth until marking. Ewes should be vaccinated annually prior to lambing. This gives the lamb temporary immunity through the ewe's milk supply, until it can be vaccinated itself.

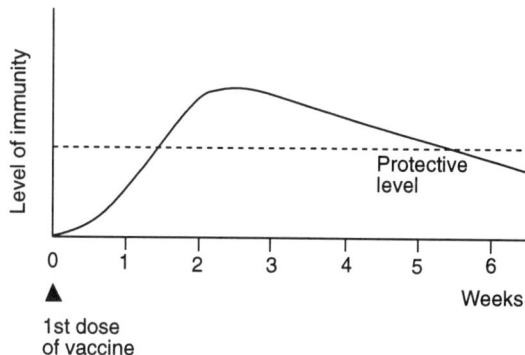

Figure 7.9 *Level of immunity after the first dose*

81

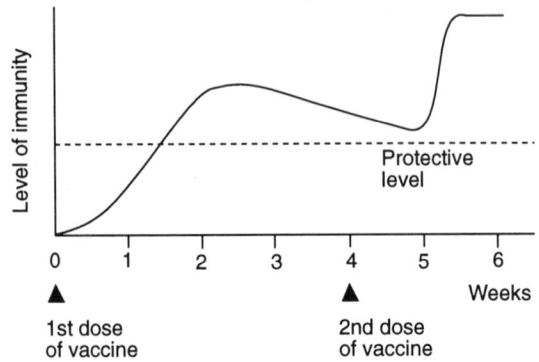

Figure 7.10 *Level of immunity after the second dose*

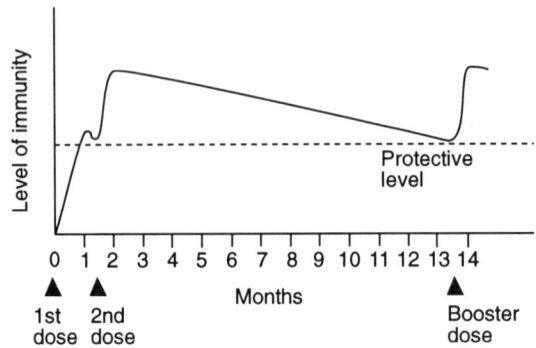

Figure 7.11 *Level of immunity after a booster dose*

Wool and skin diseases

Mycotic dermatitis (Lumpy wool)

Commonly referred to as "dermo" mycotic dermatitis is a bacterial infection of the skin and wool. It is most common in Merinos and affects mainly young sheep.

Dermatitis requires moist conditions to be active and spreads quickly throughout a mob. Infections commence with a fluid pus being produced from the skin. This dries on the wool fibres to become hard lumps. As the weather dries up the infection will usually clear leaving the lumps to dry on the wool. These lumps will move off the skin as the wool fibres grow. Lumps are often hard to see and can only be detected by feeling the wool.

Options for treatment are limited, and infected sheep should be culled. In severe cases penicillin injections will lift lumps from the skin to allow shearing but this is a very expensive option. Infected sheep should always be shorn last, as dermatitis can be spread by the shearing hand-piece. Dipping sheep with zinc sulphate mixed in the dip fluid will help prevent dermatitis.

Fleece rot

Fleece rot is a bacterial infection on the skin surface which appears as a coloured stain of the wool fibres. It is more active during warm and moist weather conditions. It occurs mainly along the back line and neck where moisture collects. It is usually green or yellow but varies through a range of other colours depending on the bacteria present.

There is no cure for fleece rot but culling of affected sheep is an effective method of eradication, because resistance to fleece rot is an inherited trait. All conformation faults of the back and shoulders and excessive body wrinkle should be culled because of the association between these faults and the development of fleece rot.

Other diseases

Footrot

Footrot is a bacterial infection which affects the feet of sheep. The bacteria causes the separation of the hard and soft part of the hoof. Early signs of footrot are lameness, a reddening of the tissue between the toes, and partial separation of the hard and soft part of the hoof. In severe cases the sheep can not graze effectively and production losses can occur from reduced wool production, lambing percentages and weight gain.

Footrot is highly contagious but is only active in warm, moist weather. The bacteria can not spread to a healthy foot and there must be some damage to admit the bacteria. This typically occurs when pasture is long and remains wet for long periods. Under these conditions, the bacteria are spread by sheep walking over the same soil or pasture as other infected sheep.

Footrot can lie dormant in pockets in the hoof for long periods until suitable conditions for its spread occur, but the bacteria can only live in the soil for a maximum of seven days, even under ideal conditions.

> **Footrot is a notifiable disease in all states of Australia. This means that once the disease has been identified, the relevant government body in your state must be notified of the disease. Extensive eradication programs are in place in an attempt to eradicate the disease.**

The first sign of footrot is usually lame sheep. The foot appears red and inflamed between the toes upon inspection. Decaying of the hoof begins at the heel of the claws and works forward to the toes. The hoof should be pared back to look for pockets of the bacteria. Footrot has a distinctly putrid smell but all hooves that smell bad may not have an aggressive strain of footrot. Badly affected hooves will often be fly-struck.

The best control of footrot is prevention:

- Check all sheep before purchase and obtain a footrot free declaration from the vendor. This allows the sheep to be returned if they are found to be infected with footrot.
- Keep purchased sheep separate from the remainder of the flock until after the first footrot spread period.

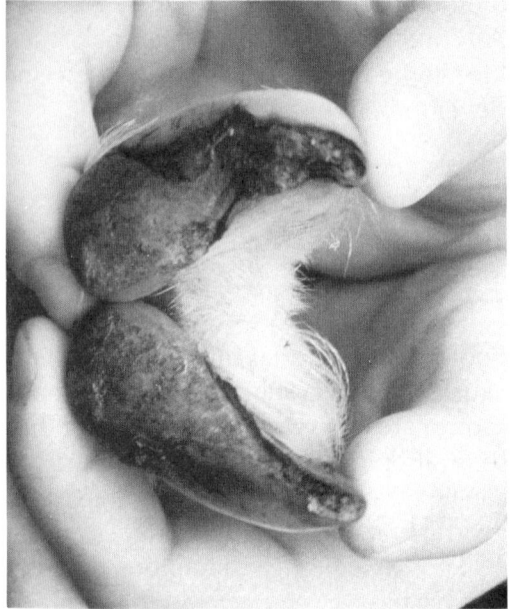

Figure 7.13 *Normal hoof showing hard and soft areas and sites for infection*

Figure 7.14 *Early stage of footrot showing reddening between the claws*

- Footbath all sheep as they are brought onto the property.
- Do not drove sheep along public stock routes unless absolutely necessary, and then treat the mob as you would a bought in mob.
- Keep boundary fences secure.
- Monitor any lame sheep and check each flock in spring and summer for signs of infection.

Eradication requires specialist knowledge to design an efficient and cost effective program. Sale of the entire flock is a sure eradication method but in most cases this is not economical and producers may want to keep breeding stock. Before beginning any treatment program, a vet should be consulted to accurately identify all the costs involved with each option and assist in developing a control strategy.

There are three options to eradicating footrot if stock are to be retained on the property:

1. Pare all feet and treat with a foot bathing solution. This may involve standing sheep in the solution for up to an hour. Inspect and cull all sheep still affected after ten days.
2. Treat all sheep with an antibiotic, then stand the sheep on grating for 12 to 24 hours. Inspect and cull all sheep still affected.
3. Do not treat animals but begin inspections and culling.

With these options, inspection and culling continues until the flock has passed two inspections. The summer period is an ideal time to conduct an eradication program, because the disease does not spread over dry summer conditions.

Footrot vaccines are available and these may be useful in the control phase of a footrot eradication program.

Foot abscess

Foot abscess is more likely in heavy animals such as rams and Crossbred sheep. Usually only a small number of stock are infected.

This disease is often confused with footrot. Foot abscess has many similar symptoms to that of footrot but is less severe and much easier to eradicate. Unlike footrot, foot abscess causes the foot to become warm and swollen. In extreme cases the abscess will burst discharging a fluid puss.

Treatment can be by injection with antibiotics, or by paring the infected foot and letting it drain.

Scabby mouth

Scabby mouth is a viral infection which causes thick scabs to form on the mouth, nose and lower legs of infected sheep. The virus is only present in certain areas and lives in the soil. It enters the sheep's body through abrasions or scratches in the skin.

Sheep build up a strong immunity to scabby mouth once they have been infected by the virus. The best form of treatment is to let the virus run its course and healing of the scabs will occur in three to four weeks.

In areas where scabby mouth occurs vaccination is recommended at lamb marking time to prevent the disease.

Brucellosis

Brucellosis is a sexually transmitted disease which causes infertility in rams. Ewes can carry and spread brucellosis but it does not affect them.

Rams suspected of having brucellosis need to be blood tested by a veterinarian and culled if tested positive. Ensure that replacement rams are only purchased from an accredited brucellosis free flock.

Cheesy gland

Cheesy gland is often referred to as "shearers boil" or "yolk boil". Cheesy gland is caused by bacteria which form large abscesses in the lymph glands of sheep. These abscesses often burst during shearing and contain a thick yellow-green pus. Cheesy gland can cause major losses with part or whole carcases being condemned at the abattoirs.

Infected animals can not be treated. Prevention can be achieved by vaccination, available in common "6 in 1" vaccines. Good hygiene practices at shearing time prevents the spread of cheesy gland.

Pizzle rot (Sheath rot)

This is a bacterial infection which affects wethers grazing on lush, predominantly clover pastures. Pizzle rot starts as small ulcers around the tip of the sheath. These develop into scabs and usually become infected. In severe cases the end of the sheath will become blocked preventing urination. The general symptoms are swelling in the pizzle area, arched back and the animal will look hollow and dull.

As a preventative measure, inject animals with testosterone in early spring prior to pasture becoming lush. If possible move sheep into paddocks with the least amount of clover.

Sheep with mild infections will recover in two to three weeks after being moved from lush, high protein pastures. In severe cases it is necessary to lance open the swollen area at the sheath opening to drain fluids. Flush the infected area with disinfectant and apply flystrike powder.

Pink eye

Pink eye is an infection of the eye which occurs during dusty and dry conditions. It will spread by close contact between animals. Flies will also spread the infection. The first signs of pink eye is swelling around the eyeball and a continual stream of tears running out of the corner of the eye down the cheek. As the condition develops the eye becomes cloudy and eventually the animal will become temporarily blind. In severe cases of small white pimple will develop in the centre of the eye.

Yarding of stock should be avoided to prevent spreading the infection once it occurs in a mob.

The best treatment is to let the infection run its course, and the eye will recover. Antibiotic powders or sprays can be administered daily to help recovery but generally this has a limited effect.

Cancer

The most common areas for cancers on sheep are the nose, ears and tail, and the vulva of ewes. This cancer is skin cancer caused by the sun. Cancer starts as a small scab which quickly develops into a large growth.

Correct tail docking and mulesing procedures can help in the prevention of tail and vulva cancer. Infected animals should be destroyed as quickly and humanely as possible.

CHAPTER 8

Husbandry operations

Stock identification

Mouthing

Mouthing is carried out to determine the age of sheep. Animals are penned into the handling race, the operator restrains the sheep and opens the lips to expose the teeth. Age can be determined by the number of permanent incisor teeth which the animal has (see Figure 8.1).

Mouthing is most commonly carried out on old sheep to determine their suitability to be kept in the flock. When sheep reach the age of five years their teeth can become loose, develop wide gaps and fall out. This is referred to as a "broken mouth", cast for age or CFA. The exact age that the teeth begin to fall out depends on the type of country that the sheep have been running on. In drier harsh country, teeth can begin to fall out as early as four years of age, but in softer country sheep's mouths can remain sound until seven to eight years. Once a sheep's mouth starts to "break" they should be culled and sold as they can not eat properly and become unproductive.

Branding

This is the identification of sheep with a station brand. The brand is dipped in branding fluid and stamped on the sheep's back. Different coloured branding fluids or branding sheep in different positions may be used to identify separate mobs. Only recommended scourable branding fluids should be used to avoid wool contamination. Branding fluid must never be diluted with other liquids as this will alter their scourable properties.

Branding is usually done after shearing but can be done when the wool is longer. If brands have been applied for less than 12 months at shearing time, or are thick and clumpy, they should be removed from the fleece.

Birth to 12 months, lamb's teeth

23–36 months, six-tooth

12–19 months, two-tooth

28–48 months, eight-tooth

18–24 months, four-tooth

Old sheep, broken mouth

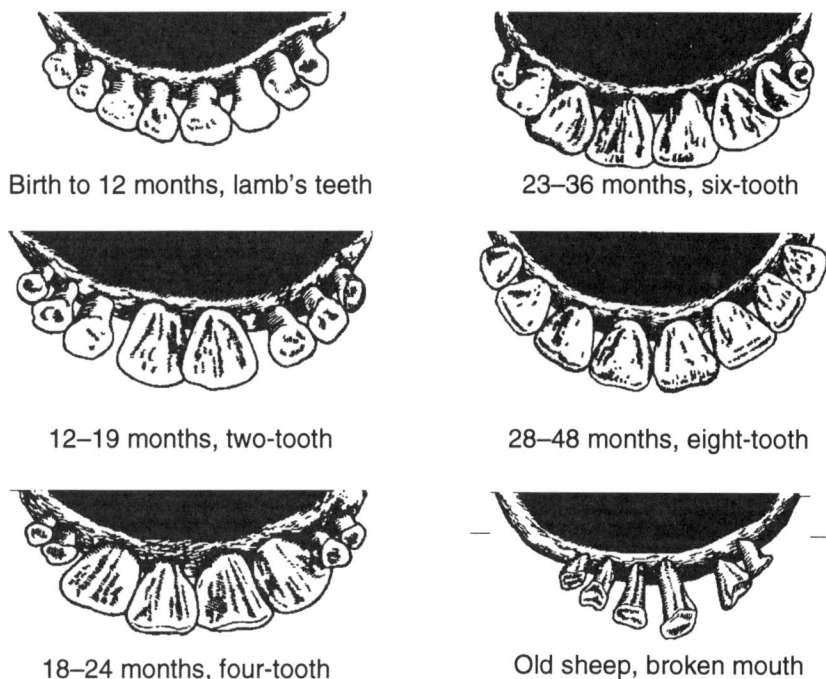

Figure 8.1 *Tooth formations which will determine a sheep's age*

Lamb marking

Lamb marking includes a number of management practices being carried out on lambs four to ten weeks of age. These include:

- tail docking
- castration
- ear marking
- ear tagging
- vaccination
- mulesing.

Ewes and lambs are yarded, and the lambs drafted off and put in a small holding pen a few at a time. Lambs are individually caught and put upside down in a cradle which holds them firm for lamb marking or are held by a catcher. Use portable yards where possible, moving the yards to a new position each day to prevent infections and to reduce dust.

Tail docking

Docking is where the lamb's tail is cut off. It is cut to a length level with the tip of the vulva in ewes, and a similar length in rams. There are three main ways of tail docking:

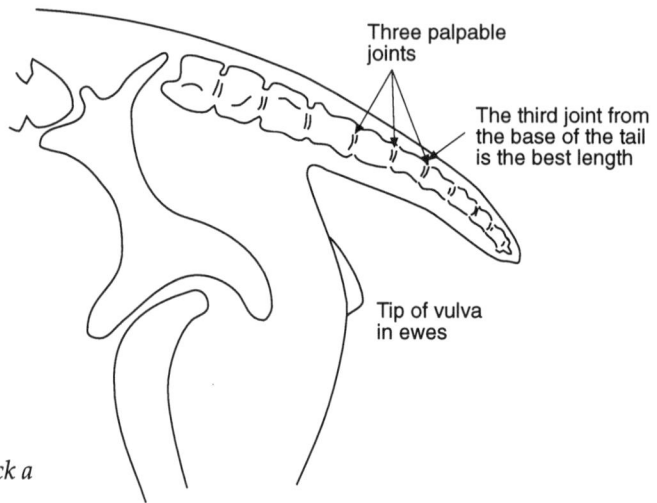

Three palpable joints

The third joint from the base of the tail is the best length

Tip of vulva in ewes

Figure 8.2 *Where to dock a lamb's tail*

1. Knife. The knife blade is placed on the tail and moved slightly towards the anus until the blade clicks in between the joints of the tail . This leaves a small flap of skin on the end of the tail to aid healing. The tail is then cut off by pressing down on the knife and pulling up on the end of the tail at the same time. In conditions susceptible to flystrike, some preventative treatment should be applied.
2. Elastrator rings. These are small rubber rings which are put onto an applicator, which stretches the rings open and allows them to be placed

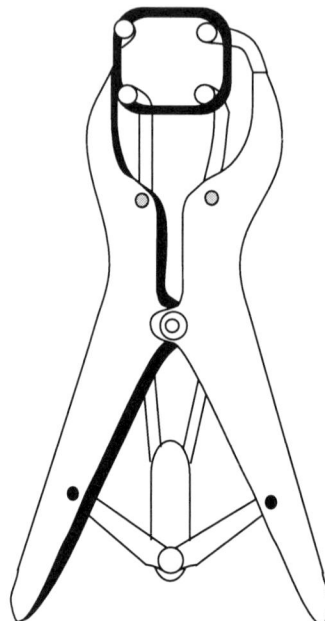

Figure 8.3 *An Elastorator ring applicator for tail docking*

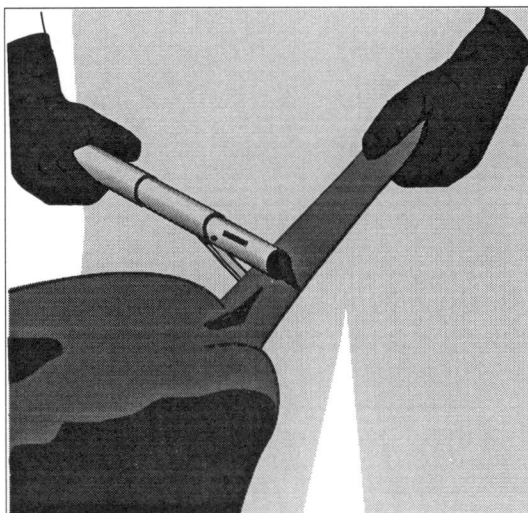

Figure 8.4 *Hot knife tail docking*

over the tail. With the ring in position, the applicator is removed and the ring closes tightly around the tail cutting off blood circulation. The tail drops off in approximately three weeks. This method can attract blowflies, especially in warm weather, so it is recommended to apply fly treatment when the ring is applied (Figure 8.3).

3. Hot knife. This is a blunt, gas heated blade. The blade clamps over the tail at the desired length and the combination of heat and pressure removes the tail. This operation is bloodless, since the blood vessels are sealed by the heat of the blade (Figure 8.4).

Castration

This is the removal of the testicles and can be done using either knife or Elastorator rings.

The lamb marking knife is a specially designed knife with a blade at one end and either a clamp or hook at the other. The end of the scrotum is cut off, and the base of the scrotum is gently squeezed between the thumb and forefinger. This will cause the tips of both testicles to protrude. If using a hook knife slide the hook down over the first testicle until it is around the base. Hold the testicle against the hook with your thumb and carefully and firmly pull the testicle out. Repeat the same procedure on the second testicle. A clamp knife is used in a similar manner, except it clamps onto the tip of each protruding testicle which are pulled out slowly. It is important to only cut the very tip off the scrotum; a rupture of the abdomen can occur if the cut is too far down the scrotum.

Elastrator rings are the same rubber rings used to remove tails. The ring is loaded onto the applicator and stretched down over the scrotum. Still holding the ring open place the thumb and forefinger around the base of

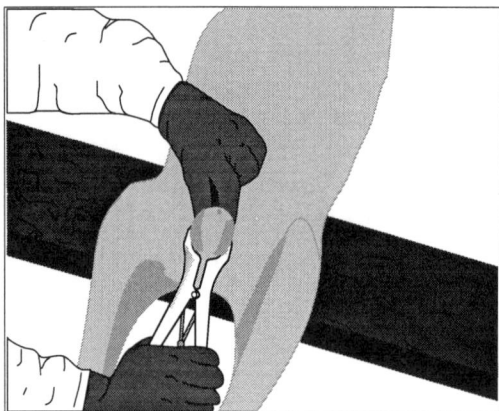

Figure 8.5 *Castrating male lambs using an Elastorator ring*

the scrotum and squeeze the testicles up into the scrotum. While holding them in place remove the applicator and check that both testicles are still in the scrotum, and that the rubber ring is not around the teats. The scrotum will drop off in approximately three weeks. Using rings is bloodless and there is less risk of infection, but recovery from castration using rings is slower and there is a greater risk of flystrike.

Earmarking

This is a mark cut into the sheep's ear as a permanent means of identification. In each state a departmental body is responsible for issuing landholders with a registered mark to identify their sheep, for instance the Rural Lands Protection Board in New South Wales. Earmarking pliers can be ordered from most rural supplies shops.

The mark is cut into a pair of specially made pliers, which fit over the ear and cut the symbol out of the ear. The earmark is put in the left or right ear depending on the sex of the sheep. This side varies from state to state: in Queensland, females are marked in the left ear and males in the right, opposite ears are used for each sex in New South Wales.

It is also common for producers to use their own earmarking system in the ear opposite to the registered earmark, to designate the age of the sheep.

Figure 8.6 *Correct positions to earmark and ear tag*

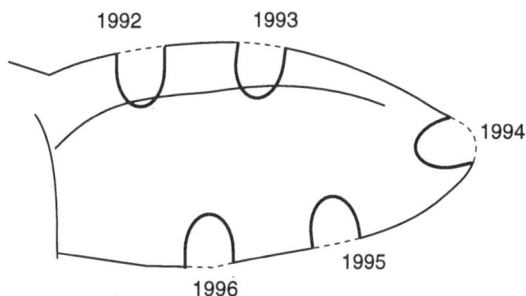

Figure 8.7 *One system to mark ears for the age of the sheep*

Ear tagging

Ear tagging is another system used to identify the age of sheep or to identify individual sheep. The tag is colour coded according to the year the lamb was "dropped" (born). Property names can also be printed on these tags at the time of manufacture. This system is not as permanent as earmarking because the tags can be caught on fences and pulled out or accidentally cut out at shearing time.

A small plastic tag is pierced through the sheep's ear with an applicator. The tag should be applied away from the edge of the ear to prevent the tag pulling out and should be between the major blood vessels seen in the ear.

Vaccination

Vaccines are given by injection. This prevents disease, by stimulating the sheep's immune system. The vaccine is injected under the skin with a vaccinating gun, taking care not to enter muscle tissue or pierce both layers of pinched up skin. The recommended positions for injection are in any non-woolled area, avoiding areas of high value in the carcase. Suitable sites include:

- the skin below the ear,
- the skin near the cheek bone,
- the bare area along the brisket.

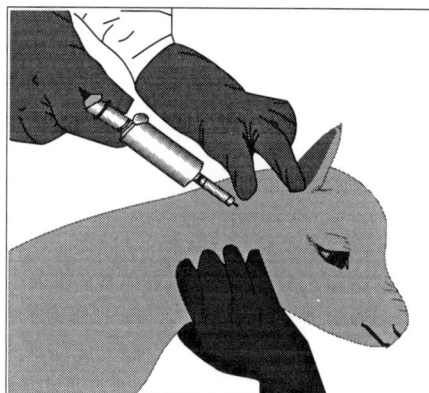

Figure 8.8 *Vaccinate lambs under the loose skin at the base of the ear*

93

The gun automatically refills from a back-pack after each dose.

Hygiene is extremely important when vaccinating and all equipment should be sterilised before and after the procedure. Equipment can be sterilised by boiling for ten minutes or using a chemical steriliser.

The needle size recommended for sheep is 18 gauge, and 10 millimetres long. Needles should be sharp and replaced after every 50 to 100 sheep. Needles should be stored in methylated spirits.

The vaccine should be kept cool at all times and manufacturer's expiry dates closely observed.

Scabby mouth treatment

A live vaccine is used to prevent scabby mouth. It is usually administered at lamb marking time in conjunction with the operations above.

The two part vaccine is mixed together, and applied using the small sharp pointy scraper supplied. It is scratched onto the bare skin inside the leg of the sheep. The point of the applicator must break the skin to allow the vaccine to enter the blood stream. This will give the lamb a mild dose of scabby mouth, but will eventually make it immune to the disease.

Other chemical applications

Drenching

Drenching is the administration of a liquid into the mouth of sheep. This is most commonly for the control of internal parasites, but can also be for external parasites and vitamin and mineral deficiencies.

The instrument used to administer the drench is the drench gun. It is a multiple dose gun with a hose connected to a back-pack carried on the operator's back. As the handle on the drench gun is squeezed, it releases a set dose of drench. When the handle is released the gun refills.

It is very important to thoroughly clean drench guns after use and leave the cylinder lubricated with a vegetable oil, to prevent the gun sticking and stop rubber components from perishing. A kit is available to replace the valves, springs, O-rings and rubber buckets, and will make an old gun function as well as a new one.

Dose

Correct dose rate is the most important part of drenching. All dose rates are calculated according to the liveweight of the animal to be drenched. For example, if a ewe would normally weigh 50 kilograms but due to dry conditions only weighs 35 kilograms, then the weight used for drench dose rate calculation is 35 kilograms. Dose rates are always marked on the drench container and manufacturer's recommendations should always be carefully followed.

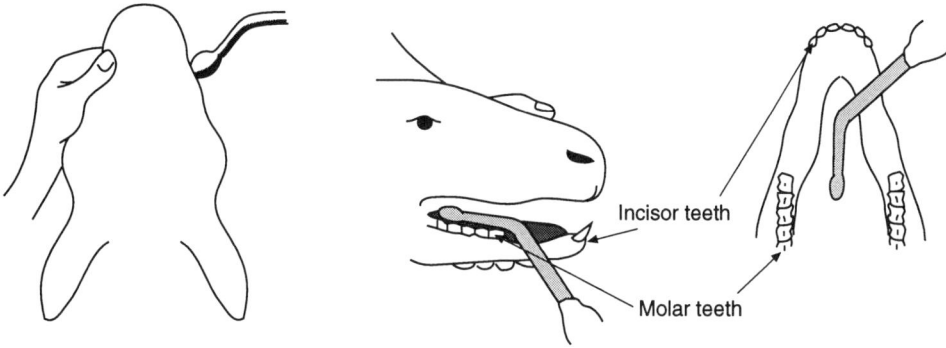

Figure 8.9 *Positioning the drench gun*

Administering drench

Sheep should be penned firmly in a race. If the race is longer than 12 metres a small gate should be used in the centre dividing the race to prevent sheep becoming overcrowded and collapsing. Some races are designed so that the operator can work from the outside. In most cases the operator needs to work from inside the race to eliminate sheep movement.

Starting from the back of the race, hold each sheep be placing one knee behind the animal and the other firmly into the shoulder. Grasp the sheep by the muzzle lifting up the top lip at the side of the jaw with your thumb. Place the drench gun nozzle into the side of the mouth and down the throat being careful not to damage the gums. Once the gun is in position squeeze the handle to release the drench. As the sheep swallows remove the gun and move onto the next animal. Alternatively, sheep can be drenched from the front. Always treat several animals before moving and use a previously treated sheep to hold animals next in line.

Jetting

Jetting is the application of a chemical onto the skin and wool of sheep to prevent blowfly strike or suppress a lice infestation. Jetting is usually intended to prevent flystrike on certain parts of the body, which is where most of the jetting fluid is concentrated, for instance the shoulder for body strike.

Hand jetting is carried out by putting the sheep into a handling race. The operator uses a jetting wand to apply the chemical. This is a T-shaped applicator with a control valve in the handle. There are several jets across the top of the T that release the chemical as the operator closes the handle. It is important that the chemical thoroughly wets the wool and skin.

Jetting races are now widely used as they are more efficient than hand jetting allowing up to 1000 sheep per hour to be treated. A jetting race is a small cage which the sheep run through. It has nozzles placed around its four internal sides and is connected to a high pressure pump. As the sheep

Figure 8.10 *Sheep jetting handpiece (NJ Phillips Pty Ltd)*

run through the cage a triggering mechanism is set off which activates the jets, and the sheep receives a high pressure blast of chemical as it runs through. However, the protection achieved by using a jetting race is less than that from hand jetting because sheep vary in the speed they travel through the jetter, and so receive differing amounts of chemical.

Jetting fluids

There are many types of jetting fluids ranging from organophosphate based products, which give from six to eight weeks' fly protection, to products like Vetrazin, which can give up to three months' protection. Some products give fly protection only, while others will also kill lice and maggots. When deciding on the chemical for a particular application it is important to carefully follow the manufacturer's guidelines and recommendations to ensure the jetting fluid is effective against the parasites being treated.

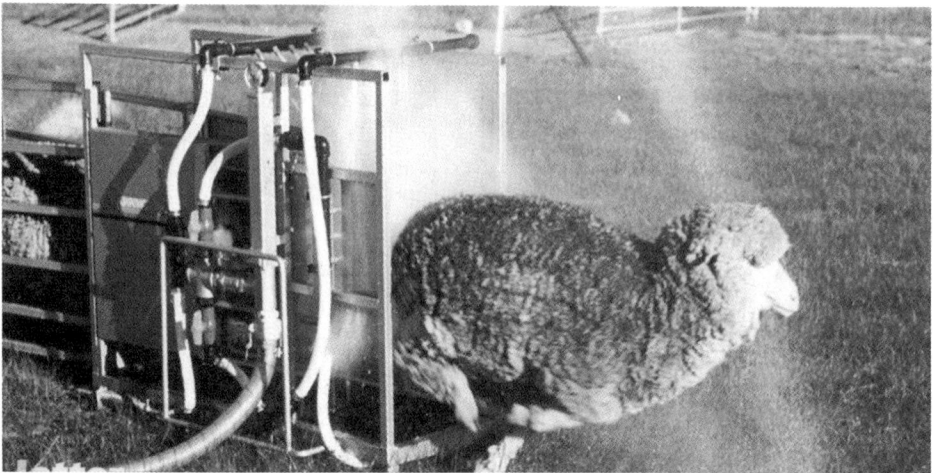

Figure 8.11 *Jetting race with sheep exiting*

Dipping

This is the application of chemical to eradicate external parasites. The chemical can be in the form of a pour-on, or it can be applied as a dipping fluid which thoroughly wets the skin.

Dipping fluid is applied in one of three ways:

1. Spray or shower dip. This is a round or rectangular pen which is enclosed with galvanised sheeting and has a series of spray nozzles at the top and bottom. These nozzles are supplied with fluid by a high volume pump. Sheep are penned into the shower dip and left with the nozzles on until thoroughly wet. This type of dipping is normally carried out four to six weeks after shearing, which allows several millimetres of wool growth to help hold more chemical. It is recommended that all sheep on a property are shorn and dipped at the same time each year. This prevents clean sheep from being reinfected by lousy sheep that have not been dipped. Lambs that are not shorn at shearing time will also need to be treated, as they may carry lice.

2. Plunge dip. This is a long narrow bath 1.2 to 1.5 metres deep. It has a race leading up to it to allow entry of the sheep. The dip is filled with dipping fluid and the sheep are forced in. As the sheep swim to the other end they are pushed under by an operator with a pole to ensure they are thoroughly wet. There is a ramp at the other end to allow the sheep to walk out into a draining pen. This method is very labour intensive as it is hard to push the sheep into the dip. Most areas have

Figure 8.12 *Circular shower dip in operation*

dipping contractors that use a portable plunge dip and offer an efficient and economical service. Manufacturer's recommendations and safety precautions should be observed as with shower dipping.

3. Pour-ons or backline dips. These treatments are applied with an applicator (similar to a drench gun) in a long strip along the backline of the sheep. The application technique is critical: the dose must be at the recommended rate, and applied evenly from the poll to the breech and equally on each side over the midline of the back, to ensure proper distribution of the product.

Only sheep that are cleanly shorn should be treated with backlines, and should be treated within 24 hours of shearing. Sheep with a bad infestation of lice or some other problem like dermatitis may not be cleanly shorn and an alternative treatment should be used. Backline products can take up to six weeks after treatment to eradicate lice. Ewes must be shorn and treated at least six weeks before lambing to prevent lice infesting the lambs and then being transferred back to the ewes. If this initial treatment is applied correctly then a further treatment will not be necessary until after the next shearing in 12 months' time.

SAFETY

Manufacturer's instructions should be carefully read and followed when dipping, to ensure chemicals are the correct strength and that "reinforcing" is carried out as recommended.

Safety precautions should always be followed as the chemicals are readily absorbed through the skin or inhaled. Always wear gloves when handling these chemicals.

There are three main types of products used to control external parasites:

1. Organophosphates: applied using a shower or plunge dip. There is little pest resistance to these chemicals evident in Australia.
2. Synthetic pyrethrins: applied using a shower or plunge dip or as a backline treatment. There is resistance to this group of products recorded in Australia. Many producers use backlines for one or more years and then change to another dipping program for a year to slow the development of resistance.
3. Insect growth inhibitors: the new generation of dips. These are applied as a plunge or shower dip and work by interfering with the development of the insects outside shell.

Physical operations

Crutching, wigging and ringing

These operations are carried out when sheep have six to seven months' wool growth. Crutching is to prevent the wool from the breech area being stained with urine and faeces. This keeps the wool cleaner at shearing time and prevents blowfly strike. The sheep is sat on its rump and a sweep (blow) made with the shearing handpiece around the inside of the legs. The sheep is then rolled to one side and the wool is shorn from around the breech area. Care should be taken not to damage the teats, cut the vulva or sever the hamstring. Crutching ewes can include one blow in front of the udder. A very light crutch or "bung hole" can be used to clean up dirty sheep prior to sale or slaughter.

Wigging involves shearing wool from around the eyes of sheep to prevent them becoming "wool blind", and to prevent grass seeds getting into the eyes. Wigging is usually done in conjunction with crutching.

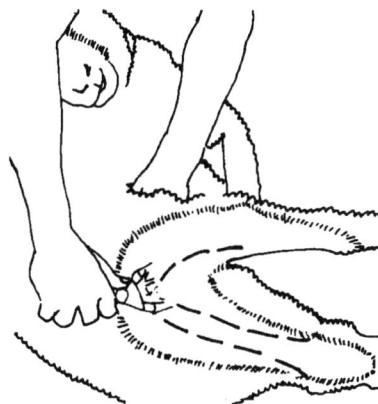

Figure 8.13 *Crutching a sheep to minimise the threat of fly strike*

Figure 8.14 *Wigging to prevent wool blindness*

99

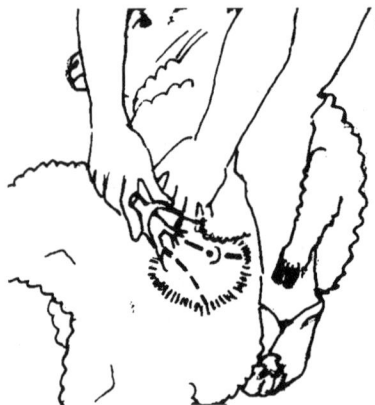

Figure 8.15 *Ringing*

Ringing is the removal of wool from around the pizzle of rams and wethers. Ringing is carried out to prevent fly strike and to minimise stained wool at shearing time. This operation is also carried out with crutching.

Mulesing

Mulesing can be done at lamb marking time or at a later date, but usually before the lamb is 12 months old.

The operation is carried out with the lamb being held in a lamb marking cradle, using a sharp pair of shears. The aim is to cut away the woolled skin from around the tail and down either side of the anus and vulva, tapering off the cut as it runs to a point about half way between the tail and hock. A small "V" of untouched skin should be left over the tail, as this helps to prevent skin cancer.

Mulesing works on the principle that if the woolled skin is cut away the bare skin next to the mules will stretch as healing takes place. This creates more bare skin around the anus and vulva which reduces wool stain and in turn reduces fly strike. Some producers prefer a light mules whilst others use a heavy mules known as a "radical mules".

Mulesing powder should be applied to the wound after mulesing.

Blowfly treatment

When sheep are struck by blowflies they need to be treated immediately to prevent the spread of maggots. Shear or clip the wool away from the infected area, taking care not to miss any small pockets of maggots around the infected area. When all of the wool has been shorn away, exposing the maggots, the infected area should be thoroughly soaked with fly dressing to kill the maggots. It is also recommended to treat the wool around the infected area with fly dressing to prevent any further strike.

Breech cuts

Leave a band of skin 2 cm wide. This aids healing and gives maximum stretch either side of the bare area.

The width of the cut will vary with age of lamb and degree of wrinkle. The range should be about 3 cm (young firstcross lamb) to 7 cm (weaned wrinkly Merino lamb).

Finish cuts in a point less than halfway from anus to hock. Any further down will only delay healing

pin bone

natural bare area

Cuts start with a point. Place a blade of the sheers on each side of the pin bone when starting this cut.

The cuts must border the bare area but do not cut into the bare area. Healing of the wounds then widens the bare area without distorting it

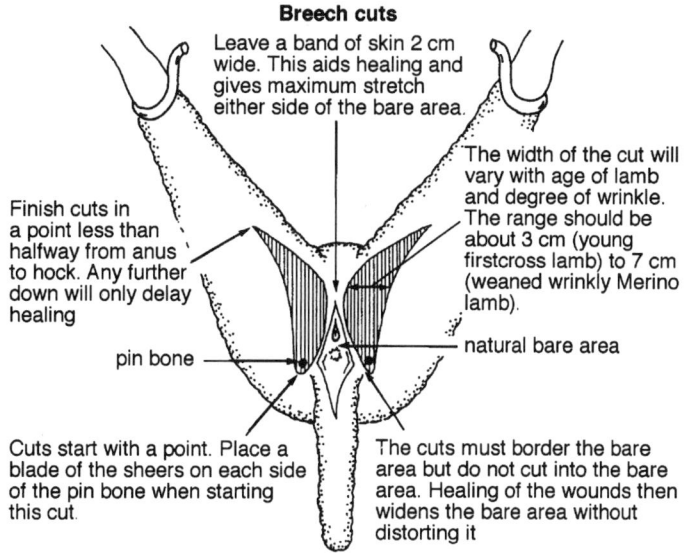

Figure 8.16 *Mulesing*

Foot paring

This is the trimming of excess hoof from sheep's feet due to one of the following conditions:

- Footrot or foot abscess. When sheep are lame one of these diseases is usually suspected. The sheep needs to be caught and turned over and the excess hoof trimmed back to expose any problems.
- Overgrown hooves. In country where pasture conditions are soft and lush, sheep's hooves become long and overgrown. Annual trimming is required to keep feet in good order.

Trimming can be done by hand operated foot shears, or by air operated mechanical shears. Commence at the heel of the hoof and pare along the outer wall making sure the wall is flush with the sole of the foot. Trim the inner wall in the same manner with both cuts meeting at the toe.

Figure 8.17 *Well trimmed feet*

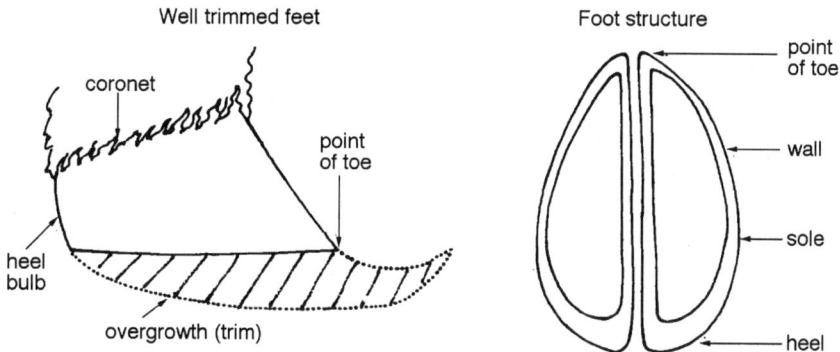

Well trimmed feet

Foot structure

coronet

point of toe

point of toe

wall

sole

heel bulb

overgrowth (trim)

heel

CHAPTER 9

Shearing

Shearing is the end process of 12 months' careful management of the flock to maximise wool production and produce a high quality clip. Income from wool is a major annual income source for many sheep producers. Every person working in the shed must strive to produce a clip that is free from contamination and stain, and prepare the clip to suit the requirements of early stage processing mills.

Contamination

Contamination costs the Australian wool industry millions of dollars each year. This cost is eventually passed back to the wool grower in the form of lower prices and reduced market share. Wool has direct competition in the market place from man-made fibres. Unlike wool, the synthetic fibre is not sold to the customer containing baling twine, towels, stains or various items of machinery.

Prevention

Contamination occurs when any non-wool material or stained wool gets into the clip. It can not be wholly prevented by quality control measures such as cleaning up around the shed, providing a few clothes hooks away from the press, and pulling out stain when it is found on the wool table. Contamination can however be prevented by a change in attitude to the wool production process. Producing a clip is a year long process and reducing contamination should be viewed the same way. The main contaminants, the problems each causes, and what can be done during the production year are listed below. Many of these involve keeping contamination in mind throughout the year rather than just at shearing time.

Clean the shed and yards prior to shearing. Remove dust, spider webs, bird nests and so on from the shed to prevent contamination. Maintain pride in the quality of work around the shed: workers will not feel compelled to do the best they can in dirty and inadequate facilities.

Table 9.1 *Wool contaminants and the measures recommended to eliminate them*

Contaminant	Problem for processors	Solution
Urine stain.	Dark coloured fibres show up as a dirty mark, especially in white and light coloured fabric.	Mules all sheep. Crutch all sheep within 3 months of shearing, ensuring all stain is removed in this process.
Black and pigmented wool.	These fibres must be removed by hand once they are in the fabric.	Cull all sheep with black spots. Shear black sheep last. Remove the entire fleece from any sheep with pigmented fibres and brand as "Black".
Polypropylene, including baling twine and wool material	this material breaks up in early stage processing and is distributed throughout the batch of wool. Twine is usually coloured and this shows up in the fabric. Neutral coloured fibres like those from wool packs do not take dye.	Collect al hay baling twine as the hay is fed out to the flock and deatroy it. Only use twine for what it was designed for baling hay. Never use twine in the shed or yards for any purpose. Do not use old super bags for storing wool. Carefully shake out wool packs outside the shed and inspect for loose fibres. Only usenew woolpacks. Handle wool bales carefully to avoid tearing the pack
Clothing and towels	These break up and are shredded in processing, and are blended throughout the batch They show up as coloured fibres or dye a different colour.	Provide and use hooks for towells and clothing away from all wool handling areas
Tools, wool hooks and branding equipment	These may cause damage to wool core sampling tubes, or to processing equipment.	Keep all tools in secure storage in the expert room, and away from all wool handling areas Only have equipment needed by the shearers near the shearing stands. Store and use all branding equipment away from the wool handling area. Keep wool hooks away from the wool handling area and always return the hook to the correct place

Table 9.1 *(Continued)*

Contaminant	Problem for processors	Solution
Dog hair and animal hair	As for pigmented fibre, this shows up as a fault in the finished fabric.	Keep dogs out of the wool handling area of the shed. Tie all dogs up outside the wool area when not being used. Bale all wool at the completion of shearing to prevent animals camping or living in the wool.
Wire, nails and bale fasteners	Damage to processing equipment.	Discard used wire. Keep nails for repairs in the expert room away from the wool handling area and make sure nails in floorboards are secure. Keep bail fasteners away from the wool handling area until needed to close a butt or fasten a bale.
Cigarette butts	These break up and the filters are spread through the wool top.	Do not smoke in the shed. Provide a smoking area away from any wool handling areas. Provide ashtrays in the area.
Food and general rubbish	This causes general contamination of the clip. It also can cause damage to processing equipment, and it destroys customer confidance in the quality of the product.	Have a separate eating area in the shed. Provide adequate garbage bins in this area, at entry points to the shed and in a convenient position for shearers. Develop an awareness of cleanliness around the shed.

Shearing preparation

Sheep

It is important for sheep to be yarded at least four hours prior to being shedded to allow them to empty out. This is done to keep the sheep and shed as clean as possible, it makes shearing easier, and it is an agreed clause in the Pastoral Industry Award.

Sheep should be drafted into mobs of similar wool type to allow for better clip preparation and classing. Ewes, wethers, weaners and lambs should go through the shed in separate mobs.

Any mobs that contain sheep with stained wool around the crutch or pizzle should be either crutched prior to shearing or shorn last. Sheep that have black or pigmented fibres should not be run with the main flock. Any sheep carrying pigmented wool should be shorn last.

Sheep have to be dry for shearing. If a shearer believes the sheep are too wet to shear, shearing stops and a vote is taken. The vote decides if the sheep are let out to dry and shearing recommences the next day, or if shearing continues.

Yards

General maintenance of the sheep yards should be up to date to ensure that they work efficiently throughout shearing. Check that all gates swing and can be securely closed. Yards and the surrounding area should be clean and free from things that may contaminate the wool clip.

In dry conditions it is advisable to water yards prior to shearing, and regularly during shearing. This will minimise dust in the wool.

The shed

Maintenance in the shearing shed should be completed prior to shearing with consideration given to the following:

- Shearing machines or overhead gear serviced (including downtubes).
- Wool press serviced.
- Wool bins set up.
- All pen gates swing freely.
- Any broken floorboards or grating repaired.
- The board washed.
- The shed cleaned (removing anything that may contaminate the clip).
- Ashtrays placed in the designated smoking area.
- Supplies of woolpacks, grinding emeries, emery glue, bale fasteners, bale branding stencils and ink, oil, shearing tools, soap and fly strike treatment are adequate.
- Arrange supply of shearing stationery including a wool book, tally book and classer's specifications. (Contact the wool broker.)

Lighting must be sufficient for all shearers, shedhands and the classer to do their job as well as possible.

A separate eating area should be supplied away from the stands and wool working area. Generous rubbish bins should be provided.

Shearing shed duties

Labour

On larger properties a shearing team is employed to complete all the tasks in the shearing shed. A shearing team for a four stand shed would normally consist of the following:

- 4 shearers,
- 1 wool classer; who skirts and classes the wool,

- 1 wool roller; who helps the classer,
- 1 rouseabout; who picks up fleeces and sweeps the board,
- 1 wool presser,
- 1 cook. In places where the team stays on the property in quarters, a cook is employed to prepare meals.

In areas where sheep numbers are smaller a woolgrower will only employ individual members of the above team as needed. For larger teams, additional staff may include:

- An "expert" who pens up the sheep, grinds combs and cutters and is responsible for general maintenance of the shearing equipment.
- Additional shedhands to handle the volume of wool as the number of stands increases.

Shearers work to a set pattern of hours for each day. They work in two-hour sessions which are known as "runs". The division of the working day is as follows:

Run 1	7.30 am to 9.30 am
Smoko	9.30 am to 10.00 am
Run 2	10.00 am to noon
Lunch	Noon to 1.00 pm
Run 3	1.00 pm to 3.00 pm
Smoko	3.00 pm to 3.30 pm
Run 4	3.30 pm to 5.30 pm.

Wool types

It is necessary for shed staff to have a clear understanding of the different categories of oddment wool (secondary wool) in order for them to be kept separate and correctly prepared. "Locks"are short pieces of wool under 15 millimetres in length, including second cuts, which are swept from the board or from under the wool table. "Pieces" come from around the edge of the fleece and are distinctly shorter than the main fleece. They are higher in vegetable matter and contain sweaty locks. "Bellies" are the wool shorn from the belly of the sheep. "Stain" is wool that is stained by urine or faeces. "Dags" are wool matted with lumps of faeces from around the crutch area.

Penning up

Penning up is the procedure where sheep are moved through the shed from pen to pen eventually ending up in the shearers' catching pens. Each shearer's catching pen is filled between runs or when there is only two sheep left in the pen. Penning up should be done causing as little stress to the sheep as possible with care taken not to pen sheep too tightly. A good dog is very useful, but should be muzzled to avoid biting.

Sweeping

The shearer's stand is swept after each sheep is shorn. The wool swept from the board is mostly locks but can contain pieces, belly wool and stained wool. The shed hand must carefully sort this wool into each appropriate oddment line.

Bellies

The belly wool will be shorn from each sheep first and thrown to the side. It is the responsibility of the shed hand to collect each of these from the board.

There are two areas of stain that may be present on the belly wool. The brisket area of the belly wool on many sheep will be short and often carrying very heavy yellow colour. Bellies from wethers may have a urine stained area of wool around the pizzle. Both areas of stain must be removed from the belly and placed in the stain line.

Removing crutch wool

The short wool around the inside the back legs will be shorn next, followed by the first hind leg and the crutch area. The wool from inside the rear legs and around the crutch may contain urine or faeces stained wool that should be removed from the locks line. It is easier to remove this wool as it is shorn rather than sort through the locks. The crutch wool can be removed with care by hand or using a sweeping "paddle".

Picking up

After the sheep is shorn the shed hand picks up the fleece and throws it onto the wool table.

Throwing a fleece takes some time to perfect. When the shearer releases the sheep, roll the left side of the fleece around until the back leg of the fleece can be seen. Straighten the back leg out. Holding the end of each back leg, push the back area of the fleece together. Still holding each leg, fold the fleece back toward the neck. Wrap the back legs under the fleece, and pick it up.

Keep the elbows tucked tightly together around the fleece to prevent parts of it dropping as it is carried to the wool table. Throw the fleece up and out onto the table, still holding the back legs. The fleece should now be spread out along the full length of the table, tip of the wool up. Be careful not to catch the front of the fleece on the table as it is thrown.

Lambs' wool is usually too short to throw onto the table in the same manner as a longer fleece. The wool table can be covered with new wool packs (after checking for the presence of any contaminating material) to prevent the shorter wool falling through to the floor.

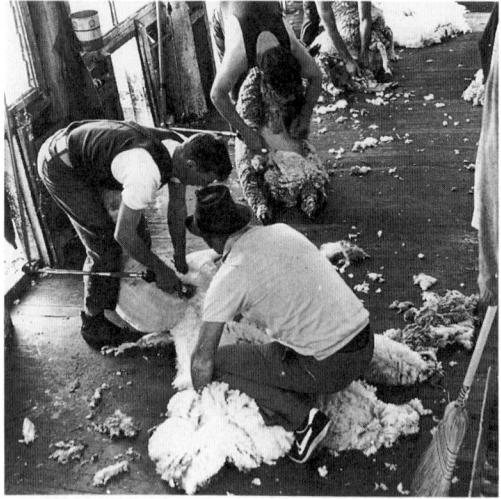

Figure 9.1 *Picking up a fleece from the board*

Figure 9.2 *Fleeces need to be thrown correctly to allow fast and efficient skirting*

The bellies and the crutch area should be removed as for adult sheep. The remainder of the fleece can be picked up by hand or using lamb batts (boards to make picking up the loose wool easier) and lobbed gently onto the wool table. The sweaty edges, shorter wool and skin pieces are removed and the remaining wool classed.

Wool classer

The wool classer is in charge of all wool operations and wool handling staff in the shed. The classer must ensure that all duties are performed correctly so that the clip is prepared to the highest possible standard. After skirting, the wool classer will class the fleece into lines according to spinning quality, length and any defects it might contain.

The wool classer needs to label wool bins clearly so that all shed staff are aware of the lines, and that wool is placed into the correct bins.

Skirting

Skirting is carried out by a wool classer and wool roller. This is the operation where the shorter wool, wool much higher in vegetable matter and fribs are removed from the edges of the fleece. Any backs where the dust has penetrated much deeper into the fleece and areas of clumpy twigs in the neck area, are also removed whilst skirting.

The wool classer determines how heavily to skirt the fleeces. This decision is critical. If skirting is too heavy, valuable fleece wool will be discounted when sold in the pieces line. Alternatively skirting too lightly results in fleece lines being discounted because all fribs and short wool were not removed.

There should be no stained wool present on the table if the sheep have been prepared to provide a quality clip. If some stain is present, this should be removed on the board. If stain is present in the crutch area of the fleece on the table, this portion of the fleece should be carefully checked. Remember, if stain is found on the table, there will probably be some stain that has escaped detection and ended up in the locks, pieces or the fleece wool. A review of sheep preparation standards and shed practices may need to be carried out for future clips.

After skirting, the fleece should be rolled up and checked for any skin pieces. These should be removed and placed in a separate line.

Pressing

The wool presser is responsible for pressing, branding and weighing bales. The presser must shake out all wool packs outside the shed before placing them in the wool press. The area around the press must be kept clean.

Figure 9.3 *Wool being pressed*

109

The presser needs to communicate with the classer regarding what lines of wool to press and must ensure that the correct wool goes into the bales.

Bales should be branded on the top (head) with the description of wool, property name and bale number before they are removed from the press. This reduces the possibility of incorrect branding. Details should be entered in the wool book at this time.

After each bale is removed from the press, it is branded on the face with the owner's brand or property name, wool description, bale number, and the wool classer's registered stencil.

Normal lines in a medium wool Merino clip would be:

AAA   Main fleece line containing the bulk of the wool.
BBB   Fleece line containing wool distinctly broader than the AAA line.
AA   Fleece line containing shorter fleeces.
AAAE Fleece line containing tender fleeces.
AAAC Fleece line containing coloured fleeces.
PCS   Pieces; skirtings from all fleece lines.
BLS   Bellies with stain removed.
LKS   Locks (sweepings from the board and under the wool table).
STN   Stained wool.

The bales are then weighed and these weights recorded in the wool book. Bale weights must be between 110 and 204 kilograms. Recommended bale weight is 195 kilograms to allow for variation in wool press scales and moisture absorbed by the wool. Overweight bales are adjusted by the broker at the grower's expense.

CHAPTER 10

Marketing stock

Marketing is one of the main areas of the farming operation where there is potential to increase profitability, by carefully planning a selling program. It is important for producers to keep up to date with market trends, and changes in processors' and exporters' requirements. These requirements are becoming more and more demanding as local and export markets become increasingly competitive. Market specifications are becoming tighter and harder to meet. It is imperative that producers target a particular market when planning enterprises to be run on their property. A good marketing program should be planned long before stock are ready to market.

Live animal assessment

Producers must be able to accurately assess the characteristics which describe the sheep they have for sale. The key factors which determine the suitability of an animal for a particular market are:

- age,
- liveweight,
- fat score,
- dressing percentage,
- breed,
- skins,
- health status.

Age
In most cases the age of the animal is known. If the age is not known, such as when animals have been purchased for fattening, age can be determined by mouthing. Mouthing is described in detail in Chapter 8.

Liveweight

This and fat score are the most important factors in determining the suitability of an animal for a particular market. Today, many producers have scales to weigh livestock. In most areas scales can also be hired.

Fat score

This can be estimated by feeling the amount of fat cover over the ribs on the midside of the animal. Sheep are rated on a score from 1 to 5:

1 score	These are very lightly conditioned animals. The index finger will easily fit between the rib bones on the midside.
2 score	These sheep are usually described in the industry as being in "store condition". The index finger will just fit between the rib bones.
3 score	This score is the "forward store" condition. The rib bones can just be felt with the index finger. Slight fat cover is evident on the animal.
4 score	This animal is fat and in prime condition. It is well rounded over the loins. The ribs can not be felt with the index finger.
5 score	This animal is over fat. The midside feels spongy and the rib bones can only be found using extreme pressure. Tail fat is also prominent.

Dressing percentage (Yield)

This is an estimate, expressed as a percentage, of the weight of the carcase after an animal has been dressed:

$$\text{Carcase weight} = \text{liveweight} \times \text{dressing percentage}$$

For example, a lamb with a liveweight of 42 kilograms and a dressing percentage of 46 per cent will yield a dressed carcase weight of 19.3 kilograms.

Factors that affect the dressing percentage of sheep and lambs include:

- Time off feed before weighing. Sheep lose gut fill when off feed, reducing liveweight but increasing the dressing percentage.
- Fat cover of the animal. Sheep with a higher fat score will have a higher dressing percentage.
- Skin length (wool length). Sheep with longer skins will have a slightly lower dressing percentage.
- Breed. Breeds with heavier muscling will have a higher yield.
- Type of feed. Sheep running on soft, lush pasture will have a lower dressing percentage.
- Carcase trim. Most abattoirs use an AUSMEAT standard carcase trim. If animals are to be sold over the hooks, check any variations to the AUSMEAT standard carcase.

Sheep	**Fat category**	Lambs

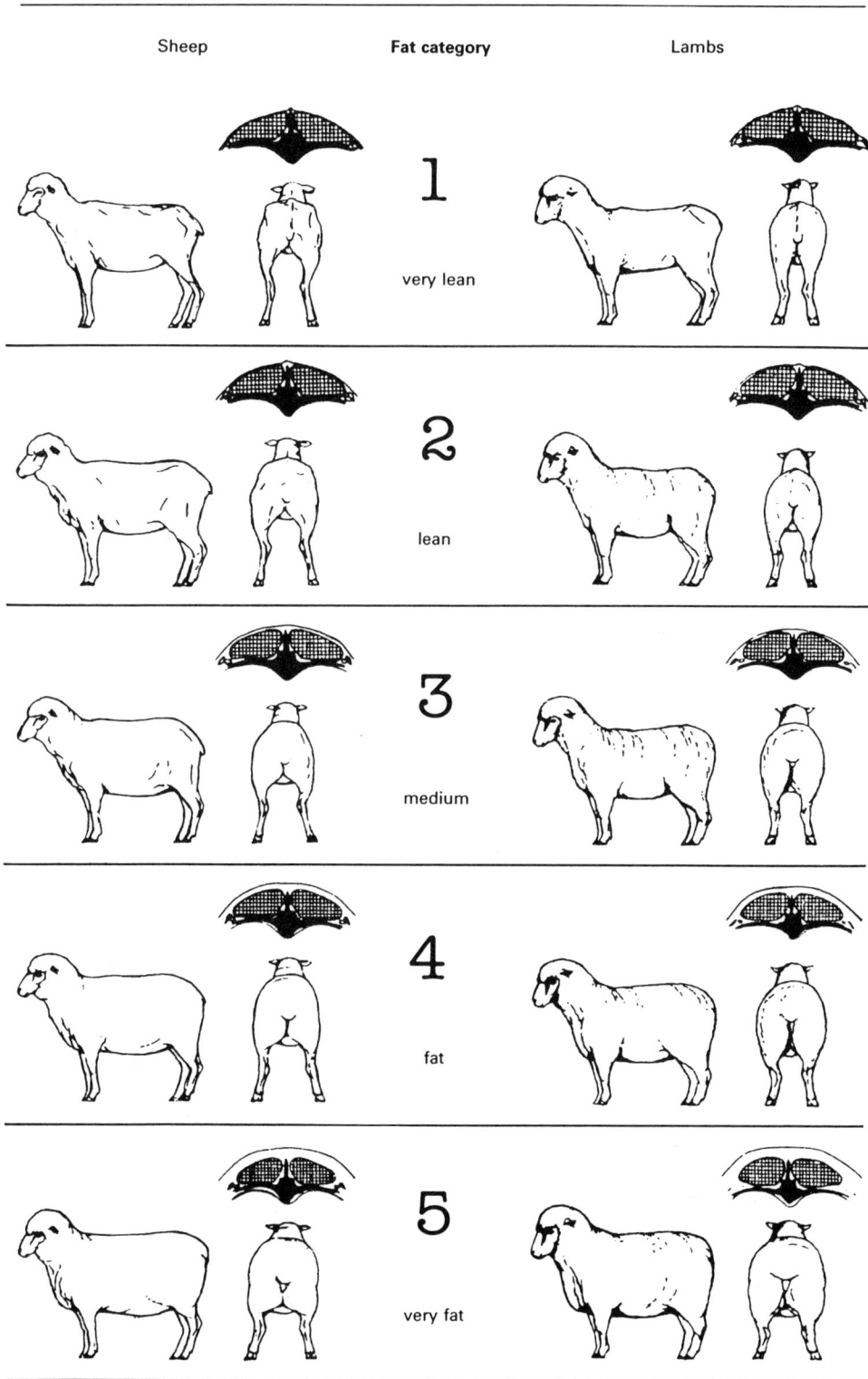

1
very lean

2
lean

3
medium

4
fat

5
very fat

Figure 10.1 *Assessing fat scores*

113

Breed

It is important to know the breed of stock, particularly with cross breeds. Market specifications often nominate a particular breed and their crosses as suitable. For example, an order might specify Merino and Merino cross lambs with a minimum of 50 per cent Merino in the cross breeds.

Skins

Skins need to be accurately described for wool length, type and quantity of burr, and dust penetration. A common description of a Merino ewe skin might be: "Wool length 70 to 80 millimetres, light clover burr on points and bellies, light dust penetration on the staple tip".

Health status

Stock need to be free from disease and parasites. Some diseases (like cheesy gland) can cause part or total condemnation of a carcase at the abattoirs. A good annual health program will ensure minimal problems in this area. Manufacturers' withholding periods on veterinary treatments also need to be observed when planning treatments for your stock and their subsequent sale.

Market specifications

Market specifications are a description of the type of animal that the purchaser wants to buy. These specifications give tolerances of certain measurable traits of a carcase that suit the buyers' requirements.

A set of specifications for live export wethers might be:
- Breed/sex   Merino wethers.
- Mouth   Guaranteed sound mouth and younger.
- Fat score   4 score.
- Weight   Average 54 kg, minimum individual 48 kg.
- Skin   Maximum length 30 mm.
- Frame   Large.
- Delivery   Port Adelaide, 15th to 20th August.

A set of specifications for local trade lambs might be:
- Breed   Second cross lamb
- Sex   Castrate male/female
- Mouth   No permanent incisor teeth (lambs' teeth)
- Fat depth   4 mm to 8 mm
- Carcase
     weight   16 kg to 18 kg
- Skin   Maximum length 50 mm
- Standard AUSMEAT trim applied.

Once producers can competently assess animals, they are able to match their stock to the buyers' requirements.

Preparation of stock for sale

Health and presentation

Stock preparation should commence several weeks prior to sale. Ensure that all animals are free from disease and parasites. Remember also to observe manufacturers' withholding periods when applying veterinary treatments. Consideration should be given to wool length. Prices will be lower for lambs with badly presented or burry skins at sale time. In this case the lambs should be shorn six to eight weeks prior to sale. Crutching should be considered if sheep are dirty or daggy as this will create hygiene problems at the abattoirs and reduce skin values. Buyers at the saleyards will heavily discount stock that require crutching.

Drafting and loading

In general everything possible needs to be done so that stock look their very best when presented to buyers at the saleyards or when delivered to the abattoirs. The following measures should be considered:

- Stock should be handled steadily and quietly during mustering and drafting to minimise stress and bruising.
- Sheep and lambs need to be yarded six to eight hours prior to trucking to empty out.
- Ensure that carriers' trucks are clean.
- Dogs should be muzzled to prevent biting.
- Do not catch lambs by the wool as this causes bruising.

Figure 10.2 *Lambs being unloaded prior to sale*

Selling methods

Selling stock through the auction system at saleyards is still the most popular method of sale. This is the best option for many producers who require a fast and simple selling method. The alternatives to auction are becoming more popular as producers look for a means of increasing profitability of their sheep enterprises. Each selling method has its advantages and disadvantages and producers should consider these carefully to select the selling system that best meets their requirements.

Saleyard auction

Stock are delivered to the local saleyards on the day of sale. They are consigned to a livestock agent who handles the drafting, penning, selling, delivery and payment for the stock. The sheep and lambs are sold on a dollars per head basis in pen lots. The agent charges a commission (percentage of sale price) for his duties. In most states the agent guarantees payment for the stock should the buyer default. This is known as acting "del credere".

The advantages of saleyard auction are:

- Fast and easy method of selling.
- Creates good competition between buyers.
- A reserve price can be put on the stock.
- Payment is guaranteed if the agent acts *del credere*.

Disadvantages are that:

- Stock are handled more than other methods.
- Length of time between farm gate and slaughter can reduce carcase quality.

Figure 10.3 *Stock being sold by auction*

- Increased freight costs if stock are passed in.
- Producers do not receive carcase feedback information.
- System is vulnerable to short term fluctuations in markets.

Private or paddock sale

Private sale involves the buyer travelling to the property to inspect and purchase the stock. The sale of the stock can be direct with the buyer, or an agent can be included in the transaction. The price offered can be a price on the farm and free of freight, or it can be a price delivered to the buyer or abattoirs.

The advantages of private sale are:

- Less handling and trucking of the stock therefore less stress and bruising.
- Since agents are not involved in the sale the producer saves commission charges.
- Producer saves freight costs if sold on farm.
- If a sale can not be made the producer has not incurred any transport or handling costs.

Disadvantages include the risk of non payment when dealing direct with a buyer. There is also limited competition between buyers, requiring the producer to know market values, and feedback information is usually not available.

Over the hooks

Over the hooks is a direct sale to an abattoir where stock are sold on a dressed weight basis. The sale price can be a set price per kilogram, but more commonly the price is determined by a grid system. This system pays a premium for carcases that meet the optimum criteria and a penalty is incurred for inferior animals. When a sale is being made under this system it is important to clarify the following points with the buyer:

- Is payment being made on a hot or cold carcase weight?
- Who is to pay the slaughter levy?
- Which carcase trim is used at the abattoir?

The advantages of over the hook sale are:

- The producer is paid a premium for quality stock.
- Feedback information is supplied from the abattoirs.
- Less stress and bruising occurs, as stock are trucked direct to the abattoirs.
- Premiums can be gained for continuous supply.

Disadvantages include penalties for stock that do not meet specifications. Again, payment is not guaranteed.

kg dressed weight	5–7	8–10	11–13	14–16	17–19	20+
18.6–19.0	1.70	1.80	1.80	1.70	1.60	–
18.0–18.5	1.75	1.90	1.90	1.80	1.70	1.50
17.5–17.9	1.75	1.90	1.90	1.80	1.70	1.50
17.0–17.4	1.70	1.80	1.80	1.70	1.60	1.50
16.0–16.9	1.50	1.60	1.60	1.60	–	–

mm fat

Figure 10.4 *Over the hooks lamb grid (All prices are in cents per kilogram dressed weight)*

Computer Aided Livestock Marketing (CALM)

This is a computerised selling system where sellers and buyers are linked for a sale. Stock are assessed on farm by an accredited CALM assessor. Their description is entered into the CALM computer system and the stock are offered for sale by auction electronically. This method offers greater flexibility as stock can be sold dollars per head, dressed weight or dressed weight on a grid. Skins can be sold with the carcase, or on a separate skin sale. The sale can be through an agent or direct with CALM.

Advantages of CALM selling are:

• Stock are still in the paddock if they do not meet the reserve.
• Payment is guaranteed by CALM.
• Transit insurance is included in the CALM fees.
• Nationwide competition on stock.
• Feedback information is available.
• There is minimum stress and bruising of stock.

Disadvantages include the assessment and cataloguing fees which still have to be paid if stock are not sold. Some buyers lack confidence in the system.

Market reports

Market reports are provided by the rural press across Australia. Many state departments of agriculture also provide an independent market reporting service, with information provided to radio and television stations. CALM market reports are available via computer, or are provided in the rural press. Many reports are also available from a variety of sources by fax.

Depending on the market involved, these reports may be in terms of cents per kilogram dressed weight plus skin value, or simply as dollars per head.

Table 10.1 *A CALM market report*

CALM MARKET REPORT

                 EASTERN STATES SLAUGHTER SHEEP/LAMBS   WEEK ENDING 14/03/97

AUCTION NUMBER:	(288)	11/03/97	TUESDAY SHEEP & LAMBS
	(289)	11/03/97	THUSDAY SHEEP & LAMBS
	(306)	12/03/97	LANDSDOWNE
			CALM SHEEP EXCHANGE

OFFERING:	22 Lot(s)	6108	Head LAMBS
	39 Lot(s)	26798	Head WETHER
	7 Lot(s)	902	Head EWES/WETHERS
	29 Lot(s)	17878	Head EWES
Total Lots	97 Lot(s)	51686	Head

STOCK TYPE	LOCATION	BREED TYPE	FAT SCORE	SKIN LENGTH	HSCW RANGE	HSCW AV	PRICE QUOTATIONS# C/KG* HSCW	$/HD
LAMBS	CENT NSW	PD/MER	2–3	1.00–1.50	16–20	17.3	204	39.20
		DH/MER	3–4	1.50–2.00	26+	26.7	192	61,41
		PD/MER	3–4	1.00–1.50	20–26	20.3	194	44.50
	SW NSW	MER	1–2	2.00–2.50	20–26	21.1	151	40.60
		BLM	3–4	2.00–2.50	20–26	24.5	213	60.20
		DH/MER	3–4	1.50–2.00	20–26	22.9	203	55.00
	W NSW	MER	2–3	1.00–1.50	16–20	19.5	163	38.20
	TAS	SUF/COR	2–3	2.00–2.50	16–20	18.5	173	43.00
		SUF/BLM	2–3	2.50–3.00	16–20	17.5	174	41.40
		TEX/BL	2–3	2.50–3.00	16–20	17.4	174	41.80
		SD/COR	2–3	1.50–2.00	16–20	17.3	180	41.60
		WSUF/CBK	2–3	2.00–2.50	–16	15.8	174	36.90
		TEX/COR	3–4	0.50–1.00	16–20	19.8	164	40.00
		SD/MER	3–4	1.00–1.50	16–20	19.0	180	42.20
		WSUF/CBK	3–4	1.50–2.00	16–20	18.8	168	42.00
		SSUF/MER	3–4	1.50–2.00	16–20	18.7	174	43.00
WETHERS	SE QLD	MER	2–3	1.00–1.50	20–26	25.6	73	25.60
		MER	2–3	0.00–0.25	20–26	24.4	65	21.40
		MER	2–3	1.00–1.50	20–26	22.9	69	22.80
	NW NSW	MER	3–4	0.50–1.00	26–32	27.9	81	28.20
	CENT NSW	MER	2–3	3.00+	20–26	25.8	99	34.00
	SW NSW	MER	2–3	1.00–1.50	20–26	25.0	82	27.40
		MER	2–3	1.50–2.00	20–26	23.6	74–76	24.90–26.40
		MER	2–3	0.25–0.50	16–20	18.0	61	17.40
	TAS	MER	3–4	1.50–2.00	26–32	27.1	39	18.20
		MER	3–4	0.25–0.50	20–26	22.0	36	14.80
EWE/WTH	TAS	MER	2–3	1.00–1.50	16–20	18.2	5.00	
		MER	2–3	0.25–0.50	16–20	17.2	4.40	
		MER	2–3	1.00–1.50	16–20	16.7	4.90	
EWES	SW NSW	MER	2–3	0.50–1.00	16–20	18.1	50	14.60
	W NSW	MER	1–2	1.50–2.00	20–26	21.0	43	17.00
	TAS	MER	2–3	0.25–0.50	16–20	16.4		3.70

*   Based on AUSMEAT Hot Standard Carcase Weight in kilograms.
    Skin value based on NSW Meat Industry Authority reports.
Price quotations are farmgate equivalent.
Source: Computer Aided Livestock Marketing (CALM) Sydney

CHAPTER 11

Marketing wool

For wool growers, marketing their clip is becoming an increasingly complex decision. More selling options are available, giving growers a wider range of marketing strategies. Alternatives, like selling wool semi-processed, forward selling and selling on the futures market, have given growers the opportunity to make better marketing decisions about their wool.

Transport

Transport is an important part of the wool marketing process. Wool sale rosters need to be consulted prior to transport to ensure the clip arrives at the store in time for pre-sale processing. The two methods of transporting wool from the farm to point of sale or delivery are road and rail. Road transport is the most common as it is the fastest. Whichever method is used, the wool must be properly covered to prevent weather damage. A copy of the classer's specifications and a consignment note must be sent with the wool or faxed to the broker prior to delivery. This allows easy identification and prompt processing at the wool store.

Most wool brokers carry an automatic sheep's back to store insurance policy which covers all wool consigned to them.

Wool testing

Most sale systems available use wool testing to let the buyer know what is being purchased. Wool testing measures the physical characteristics that determine the wool's suitability to meet a customer's processing requirements. Measurements used in the Australian wool industry have

---

ADVICE OF CORE TEST RESULTS

This Account Sale is a reproduction of details previously advised.

S23
0/00/00
0/00/00

LOT REF	NO. BALES	DESCRIPTION	NET WGT	MIC	VM	SCH DRY	ADD M/MENT MM	CV%	N/KT	POB T	M	B	COL
8C	16	AAA	3094	22.3	.9	69.5	91	18	26	50	44	6	
9C	14	AAA	2663	23.1	.8	71.1	97	16	27	46	52	2	
10C	16	AA	3126	22.8	.7	70.8	93	15	21	62	35	3	
11C	6	BBB	1170	24.8	.7	73.3	104	11	25	60	34	6	
152	12	AAAH	2291	20.7	3.4	65.3	72	15	44	60	37	3	
2002C	10	BKN	1773	21.4	4.8	56.6	77	28	25	44	50	6	
2003C	5	BLS	915	21.9	5.9	51.2	74	19	18	31	55	14	
2092	3	BKN	494	19.2	11.4	49.0	53	24	45	33	61	6	
2095	3	LK	585	20.8	5.2	49.4							
2096	4	CR	760	22.6	2.5	53.6							

Source: Elders Ltd West Wyalong (David Curry)

Figure 11.1 *Wool test results for a 90 bale clip*

been standardised internationally and are now commonly used in mills throughout the world. These measurements also allow the grower to assess the product and therefore the production methods employed.

Wool testing is carried out by the Australian Wool Testing Authority (AWTA). Two samples are taken from the bales to test the wool:

- A core sample that is used for wool testing purposes.
- A grab sample, which is displayed on the show floor on sale day, with part of it being used to test length, strength and colour.

Samples must be random and truly representative of the contents of each lot. The value of each lot of wool will be determined by these samples.

The core test sample is used to test wool for the following characteristics:

- Micron; the average fibre diameter of the sample.
- Yield; the amount of clean wool fibre in a sample. The result is expressed as a percentage.
- Vegetable matter; percentage of vegetable matter in the wool sample.

- Vegetable matter components; divided into quantities of:
  seed/shive; Barley Grass, Corkscrew Grass
  burr; clover burrs, medics
  hard heads/twigs; Bathurst Burr, Noogora Burr, Galvanised Burr.

The grab sample is used to test wool for:

- Strength; the tensile strength of wool. The result is expressed in Newtons per kilotex (N/ktex). Wool under 30 N/ktex is considered "tender". The test also gives the position of the "break" in the staple.
- Length; the average staple length of the sample in millimetres. The variation in length is also measured.
- Colour; measures the average yellowness of the sample after the wool is scoured. (The sample used to provide the micron measurement is used for this test.)

All wool sold by sample must be tested for all the characteristics provided by the core sample. Additional measurements for length, strength and colour are optional. Many mills now instruct their buyers not to purchase wool that has not been additionally measured for length and strength.

Australian Wool Exchange industry description

The Australian Wool Exchange (AWEX) has developed a system of describing the non-measured characteristics of wool. This is aimed at improving the flow of market information back to the grower.

Wool is given a "prime descriptor" of breed, wool type, style grading and vegetable matter type. "Secondary descriptors" are used to supply additional information about wool faults found in the lot, and length and strength information where wool is not measured (Figure 11.2).

For example, the AWEX description "MF4S" designates the wool as:

M   Merino
F   Fleece
4   Best Style
S   Mainly shive vegetable matter

It has no faults, and has been additionally measured for length and strength.

Wool given a description "MP5B.80W1" is:

M   Merino
P   Pieces
5   Good style
B   Mainly burr
80   Average length of 80 millimetres
W1 Slightly tender

This wool is not additionally measured, so length and strength have been visually estimated.

Selling methods

Wool may be sold by auction at a variety of sales venues, by tender, forward sold, or sold privately by the producer.

Auction

Sale by sample

Wool is consigned to a broker who handles lotting, testing, display, auction, dumping and post sale delivery. There are two types of brokers in Australia. A "flat rate" broker charges the grower a set charge per bale for his services while a "commission" broker charges a handling and storage fee, and a percentage of the gross proceeds from the sale.

When wool arrives at the store it is lotted into suitable sale lines as determined by the store wool classer. Minimum lot size is four bales and smaller lines are usually inter-lotted (combined) with other growers' wool of a similar type to make up sale lines.

The wool is then tested. The core sample is sent to the Australian Wool Testing Authority for testing and the grab sample held in security until sale. When the broker receives the test results, the wool is catalogued for sale.

On sale day, the grab sample is displayed with the test results on the show floor. Buyers inspect the wool prior to the wool being sold by public auction. After sale the wool is dumped and delivered to the buyer. Growers are paid their wool proceeds less testing, selling and handling charges, usually ten working days after the Friday following sale day.

The advantages of sale by sample are:

- Wool receives competition from companies world wide.
- Growers are paid on actual test results, not estimates.
- Growers receive test information about their wool.

The disadvantages are that:

- Growers have to wait three to five weeks after shearing for proceeds.
- Growers pay transport, handling, selling and testing costs.

Sale by separation

Wool sold by sample is stored at the same sale centre where it is displayed and sold. Wool marketed through sale by separation is stored at another centre. This may be either in country stores or another sale centre. Only the grab sample is transported for display purposes to the point of auction.

Wool delivered to Adelaide and Geelong is sold by separation in Melbourne, Brisbane wool and many New South Wales country wool brokers sell by separation in Sydney.

Reduced showing

This is a variation available in the auction system where a core sample is taken and a wool test provided, but the grab sample is replaced by a

Australian Wool Exchange Limited

AWEX-ID (NON MEASURED CHARACTERISTICS)

Version 1.3 Effective 01/01/96

| PRIME | | | | | | QUALIFIERS | | | |
| Mandatory | Where Applicable | Mandatory | Mandatory | Mandatory | Mandatory | Conditional | Conditional | Where Applicable | Where Applicable |
Breed	Wool Sub Category	Wool Category	Style FLC	VM Type	Full Stop	Greasy Length Indicator (Non SM) (F/P/B)	Strength Indicator (Non SM) (Combing)	Colour Indicator (F/P/B)	Dark Stain Indicator (P/B/C/Z)
M (Merino)	W (Weaners & Combing Lambs)	F (Fleece)	1 (Choice)	S (Seed/Shive)	• (Full Stop)	10 (6–15)	W1 (Part Tender)	H1 (Light Unscourable)	S1 (Odd Stain "D")
X (Crossbred)	L (Carding Lambs)	P (Pieces)	2 (Superior)	B (Burr)		20 (16–25)	W2 (Tender)	H2 (Medium Unscourable)	S2 (Light Stain type)
D (Downs)	Y (Black & Grey)	B (Bellies)	3 (Spinners)	N (Noogoora/Ring)		30 (26–35)	W3 (Very Tender)	H3 (Heavy Unscourable)	S3 (Meduim/Heavy Stain type)
T (Carpet)	U (Plucked & Dead)	C (Crutchings)	4 (Best)	T (Bathurst)		40 (36–45)		N (Water Stain - Green/Blue)	Q (Dags)
	K (Shorn from Skins)	Z (Locks)	5 (Good)	L (Clumpy)		50 (46–55)			
	M (Fellmongered)		6 (Average)			60 (56–65)			

			WHERE APPLICABLE		
			Cotts	Dermo	Other
O (Overgrown)	70 (66–75)		C1 (Odd Cott)	A2 (Light Dermo)	G (Doggy)
7 (Inferior)	80 (76–85)		C2 (Soft Cott type)	A2 (Medium/Heavy Dermo)	J (Jowls)
PCS/BLS	90 (86–95)		C3 (Medium/Hard Cott Type)	(Brands)	R
3 (Spinners)	100 (96–105)				
4 (Best)	110 (106–115)				D (Mud)
5 (Good)	120 (116–125)				E (Necks)
7 (Inferior)	Etc.				K (Shanks)
CRS/LKS					
1 (Good Bulk, Good Colour)					
2 (Good Bulk, Fair Colour)					
3 (Average Bulk, Fair Colour)					
4 (Inferior Bulk)					Max of two codes from this column on any one lot.

Figure 11.2 *An example of a Wool Exchange industry description chart*

proportion of the bales from the lot to be sold. Most wool sold this way is from superfine clips from the high rainfall area.

Traditional sale
Traditional sale is sale by auction, but no test is available and each bale from the lot is usually displayed. Most traditional sales are for lots where the wool is unable to be sampled to give a valid result. These include over-grown wool, unskirted fleeces, dags and black wool.

Sale by description
Sale by description systems are still being trialed. They utilise the wool testing methods mentioned above. No sample is available for physical inspection by the wool buyer. Instead the buyer relies on the standardised descriptors. This streamlines the selling of wool, by allowing electronic sales through bids entered directly by exporters, early stage processors in Australia or by overseas mills.

The disadvantages of the system are that more work is required to measure all the wool characteristics required by early stage processors, and developing buyer and vendor confidence in a sale by description system.

Sale by tender

In this system buyers submit prices by tender to the broker rather than bid on wool lots. The highest price is usually accepted. Most brokers offer a sale by tender service. Lower lines of wool (like stains or dags) are often sold this way. However some growers prefer to sell whole clips using this tender method. If the wool is passed in then it is re-offered in the next tender or auction sale as the grower requests.

The advantages of sale by tender are:
- Faster than auction.
- Sales can be continuous over several weeks and occur on any day.
- There is no testing cost.

The main disadvantage is that wool may not always receive as much competition as through auction, and prices may therefore be lower.

Private sale

Shed
This involves the wool buyer visiting the property during shearing to inspect the wool. In most cases the buyer will take samples for testing before offering a price for the wool. Growers usually get more than one buyer to make an offer on the wool and accept the highest price. Payment is immediate and cartage is the responsibility of the buyer. The advantages are:
- Growers receive immediate payment for their wool.
- Growers do not pay any transport, handling, testing or selling costs.

Disadvantages:

- Buyers discount wool to allow for fluctuations in the market.
- Shed sales may lack buyer competition if the sale does not attract sufficient buyers to the shed to offer a price.
- Growers do not receive extensive test information on their wool.
- Payment by the purchaser is not guaranteed by a third party (as is the case of sales through a broker).

Direct to the mill

This is another form of private selling. The wool is usually consigned to the mill after shearing, and tested. The mill will offer the grower a price for the wool based on the test results. The grower usually has the option to sell the wool by auction if the mill's price is not satisfactory.

The advantages of sale direct to the mill are:

- Payment is usually quick compared to auction.
- Growers can receive a premium for supplying year after year.
- Preparation of the clip will be decided in conjunction with the mill, so that the grower is assured of providing the clip as his customer requires.

Disadvantages include:

- Restricted competition on the wool if several quotes are not obtained.
- Grower may have to pay freight to the mill.

Forward selling

Forward selling allows wool to be sold up to 18 months prior to shearing. Forward selling is normally done through a wool broker or by direct contracts to mills. Details of the expected clip size, micron, vegetable matter, yield and general description of the wool type is put together using previous test results from the flock, allowing for present seasonal conditions. This information is given to the buyer, who offers a price for the clip at an agreed delivery date. There are usually allowances made in the contract to vary the agreed price if actual wool test results differ from the original description.

This option is used to even out market fluctuations. For example, if a grower expects a fall in the wool market at the time he would normally sell, then a forward contract may be considered for all or part of the clip, assuring the grower of a known return.

Advantages of this system are:

- The grower is locked into a fixed price free from market fluctuations.
- Accurate farm budgets can be formulated based on the agreed wool price.

The disadvantage is that the grower needs to monitor the wool market closely. If the wool price falls, the grower still gets the contracted price, but if the market rises, the grower has no opportunity to get a higher return.

Reading a wool market report

Market reports are provided by the Australian Wool Exchange. Trends in the wool market can be followed using the Northern, Eastern or Western Index for different sale centres. These indexes use wool market levels at July 1995 as a base.

Table 11.1 *An example of a wool market report*

AWEX–ID Closing Quotations (AUD cents/kg clean-Schlum. Dry)								
					Northern Region		Southern Region	
Micron	*Style*	*mm*	*N/kt*	*VM%*	*5-12-96*	*28-11-96*	*5-12-96*	*28-11-96*
19	MF4	90	35	1	815	764n	–	777
	MF4	80	35	1	794	748	736n	738
	MF5	90	35	1	–	–	–	–
	MF5	80	35	1	–	–	–	–
	MP5	70	35	2	659	638	–	644n
20	MF4	90	35	1	680n	671n	–	704
	MF4	80	35	1	665n	655	–	678
	MF5	90	35	1	672n	–	–	–
	MF5	80	35	1	–	642n	–	674
	MP5	70	35	2	–	559	–	552
21	MF4	90	35	1	–	634n	621	624
	MF4	80	35	1	–	622	–	–
	MP5	80	35	2	633n	612	625n	–
	MP5	80	35	2	–	520n	551	564n
22	MF4	100	35	1	–	590n	589n	590
	MF4	90	35	1	–	584	595n	–
	MF5	100	35	1	–	583n	596	589
	MF5	90	35	1	–	587	–	589
	MP5	80	35	2	–	–	536	–
23	MF4	90	35	1	–	514n	–	–
	MF5	100	35	1	–	517	512	–
	MF5	90	40	1	–	525	–	313
	MP5	90	35	1	–	516	516	–

Note: Quotes denoted 'n' are based on single lot quotations
mm = length in mm, midpoint of range
N/kt = Strength in N/ktex, midpoint of range
VM% = VMB%, end point of range

Week ending December 5, 1996

Source: Australian Wool Exchange

Butchering and dressing

Farm butchery and sheep dressing is a very simple task once the correct procedure and techniques are learnt. Slaughter of sheep and lambs for your own consumption is allowable in all states, provided that the meat and any food stuffs made from the meat are for personal consumption and are not sold.

Slaughter

* Before starting make sure that the slaughter area is clean and that knives are sharp and clean.

Figure A.1 *Butchery tools*

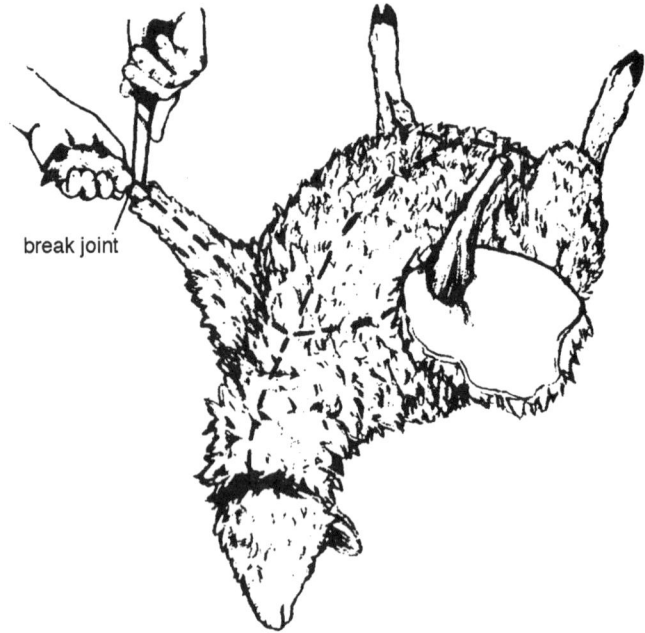

Figure A.2 *Opening cuts*

- Lock up the animals for 24 hours prior to slaughter, to allow them to empty out.
- Catch animals carefully to prevent bruising. Especially avoid handling an animal by the wool.
- Lie the animal on its side, grip the jaw and bend the head backwards around your leg. Cut the throat quickly and cleanly from ear to ear and bend the head sharply backwards to break the spinal cord. Both carotid arteries must be severed.
- Allow two to three minutes until the animal stops kicking.

Dressing

- Hold one front leg between your knees. Run the point of the knife down the inside of the leg, just under the skin, opening up the skin from the knuckle down to a point just in front of the brisket. Continue the cut from the point of the brisket down the bottom of the neck where the throat has been cut.
- Skin around the leg and down along the side of the neck. Remove the trotter at the knee joint.
- Repeat the procedure on the other front leg, making sure that the cut meets at the same point in front of the brisket.
- Trim around the oesophagus and tie a knot in it to prevent the contents of the rumen running out when the carcase is hanging up.

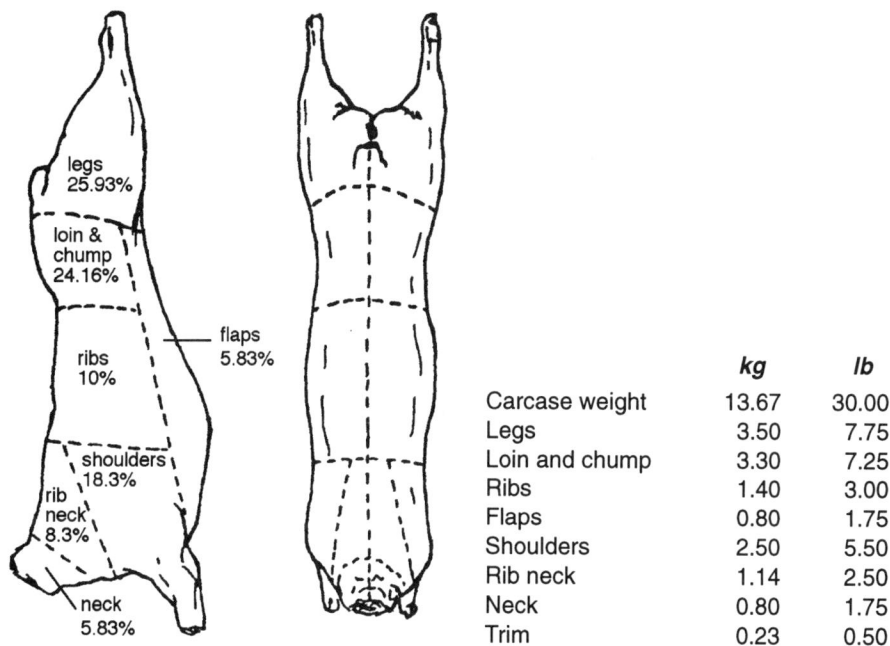

	kg	lb
Carcase weight	13.67	30.00
Legs	3.50	7.75
Loin and chump	3.30	7.25
Ribs	1.40	3.00
Flaps	0.80	1.75
Shoulders	2.50	5.50
Rib neck	1.14	2.50
Neck	0.80	1.75
Trim	0.23	0.50

Figure A.3 *Conformation and dissection of a lamb*

- Wash your hands. Placing your foot on the head, take hold of the flap of skin in front of the brisket and pull upwards, pulling the skin back off the brisket.
- Still holding the flap of skin in one hand punch under the skin back towards the pelvic area of the lamb. Punch down the sides of the ribs and the shoulders.
- Hold the rear leg between your knees. With the point of the knife open the skin down the inside of the leg to a point at the base of the tail.
- Skin back around the side of the hind leg being careful not to cut the tendon (hamstring).
- Finish skinning down the outside of the leg and remove the trotter at the last joint of the hock.
- Repeat the procedure on the opposite rear leg.
- Place the gambrel between the back legs and hang up the carcase.
- Split the skin down the centre of the belly and skin around the scrotal area out to the flank, being careful not to cut the flank.
- Wash hands. Punch from the butt of the tail down one side of the carcase being careful not to damage the selvedge layer of skin that remains on the carcase.
- Repeat on the other side of the carcase.
- Cut around the rectum and pull it out. Tie a knot in it to keep the carcase clean.

- Cut the skin from around the butt of the tail and punch the remaining skin from the hindquarter using an upward motion.
- Pull back wards on the whole skin, which will completely remove it down as far as the neck.
- Trim around the ears so that they remain on the skin, cut the skin clear from the back of the head. Cut off the head.
- Cut the stomach open from the scrotal area down to the brisket, being careful not to cut the intestines. Remove the bladder.
- Pull downwards on the large intestines and rumen and remove them.
- Remove the liver. Check, by feeling for abnormalities.
- Split the brisket and trim around the diaphragm. Remove the heart and lungs. Check for abnormalities. If any abnormalities like lumps, cysts or discolourations are found in the heart, lungs or liver the carcase should be discarded.
- Dressing is now complete. Hang the carcase in a fly- and vermin-proof area overnight prior to cutting up. It may be hung for up to one week if a coolroom is available.

Cutting up the carcase

- The carcase may be cut up once it has set, that is the meat has cooled down and rigor mortis has set in. Cutting up will be easier if the meat has set well. If conditions have not been cool enough for hanging, the carcase should be quartered and placed in a freezer for several hours prior to cutting up.
- Start by cutting off the neck with a meat saw.
- With the carcase still hanging up, using a meat saw, and starting between the back legs, cut down the centre of the backbone towards the neck until the carcase is in two halves.
- Take one half of the lamb and lay it flat on the cutting bench. Make a cut from the flank (close to the back leg) straight up to the backbone. Keep the cut at right angles to the backbone. Use a cleaver to cut through the backbone at the end of the cut, and remove the back leg from the remainder of the carcase.
- Holding the back leg about the middle, move the end of the leg to find the next joint along the leg. Cut through this joint with a knife.
- Trim the inside of the leg removing any excess fat. The leg is now ready to be frozen.
- Take the remainder of the side and hold it upside down by the front leg. Cut between the shoulder and the brisket slowly removing the shoulder from the carcase. Cut the shank off the shoulder at the same joint as the hind leg. Trim the shoulder.
- Take the remainder of the side and lay it upside down on the bench. With the meat saw make a cut lengthways along the side through the

SHEEPMEAT SIDE SKELETAL DIAGRAM

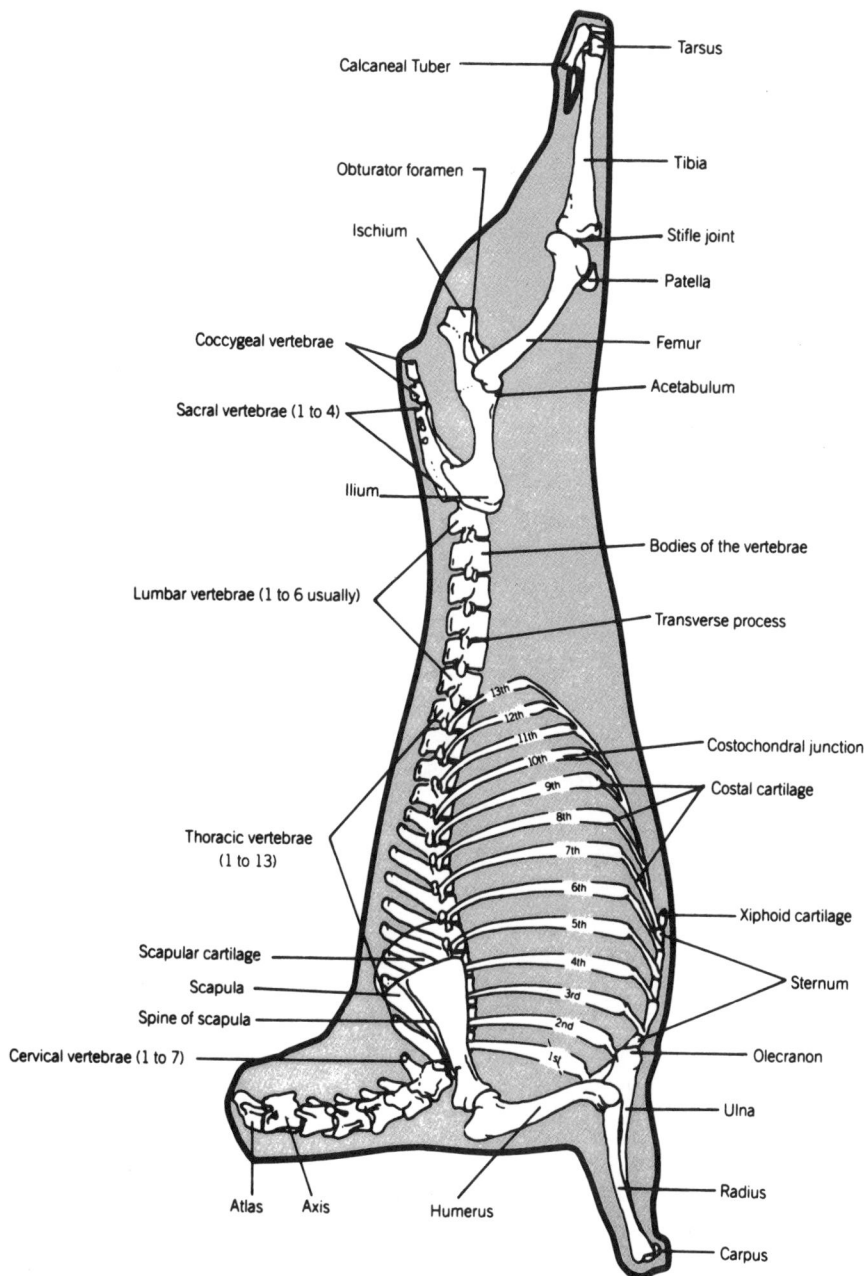

Calcaneal Tuber
Tarsus
Obturator foramen
Tibia
Ischium
Stifle joint
Patella
Coccygeal vertebrae
Femur
Acetabulum
Sacral vertebrae (1 to 4)
Ilium
Bodies of the vertebrae
Lumbar vertebrae (1 to 6 usually)
Transverse process
13th
12th
11th
10th
9th
8th
7th
6th
5th
4th
3rd
2nd
1st
Costochondral junction
Costal cartilage
Thoracic vertebrae
(1 to 13)
Xiphoid cartilage
Scapular cartilage
Scapula
Sternum
Spine of scapula
Cervical vertebrae (1 to 7)
Olecranon
Ulna
Atlas   Axis
Humerus
Radius
Carpus

Figure A.4 *Skeletal diagram of a sheep side*

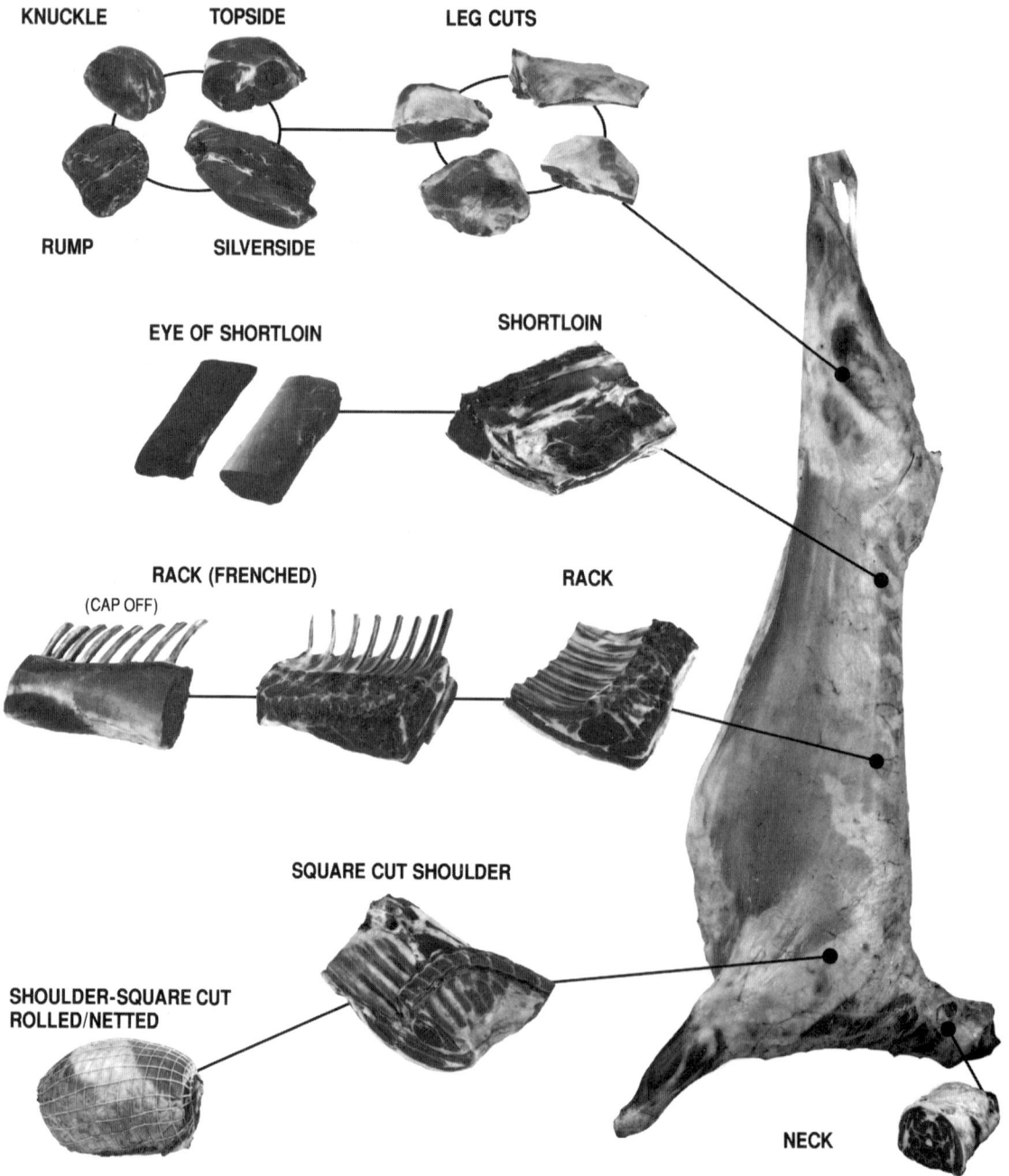

Figure A.5 *The major cuts of a side of lamb*

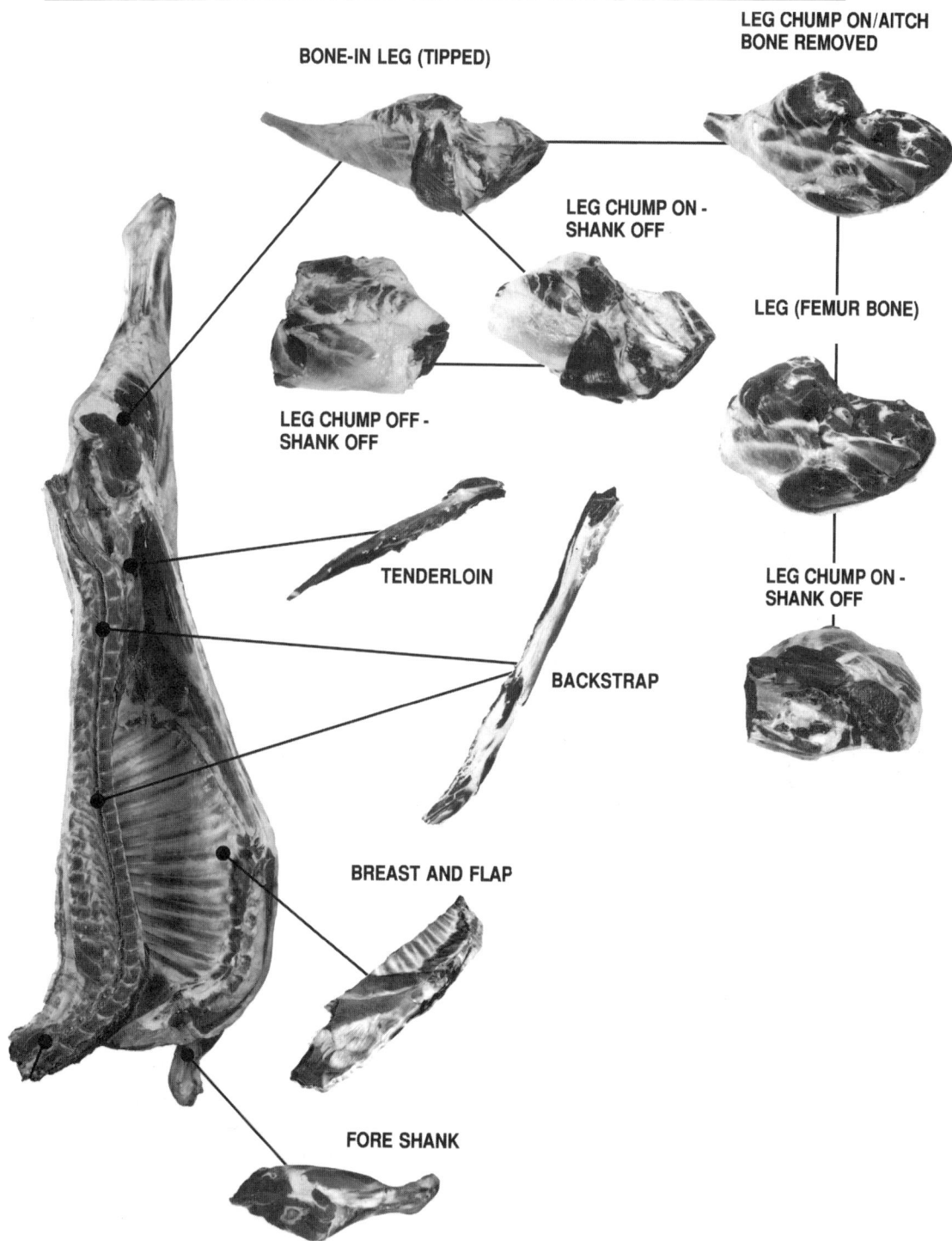

BONE-IN LEG (TIPPED)

LEG CHUMP ON/AITCH BONE REMOVED

LEG CHUMP ON - SHANK OFF

LEG (FEMUR BONE)

LEG CHUMP OFF - SHANK OFF

LEG CHUMP ON - SHANK OFF

TENDERLOIN

BACKSTRAP

BREAST AND FLAP

FORE SHANK

Figure A.5 (continued) *The major cuts of a side of lamb*

rib bones. The cut should be approximately 120 millimetres from the backbone and run parallel to it.

- The lower portion once removed is commonly called the "flap". If the lamb is not too fat the flap can be trimmed, the ends of the rib bones boned out and the flap seasoned and rolled as a roast.
- The remainder of the side can now be cut into chops. Start by finding the last rib bone towards the rear of the carcase and make a cut down along the rib bone to the back bone. Cut through the back bone with a cleaver.
- The portion of chops without rib bones are referred to as "loin" or "short loin" chops. These can now be cut to your preferred thickness with a knife, finishing the cuts through the backbone with the cleaver.
- Cut up the remaining chops by making a knife cut between each rib bone, finishing the cut with a cleaver. The last four or five "neck chops"and are only suitable for stewing.

This is the basic way of cutting up a lamb using hand tools. The butchering process is much easier if a bandsaw is available. There are many other variations to this method, the most notable being the "Trim Lamb" cutting method for heavy lean lambs. This method produces a range of cuts suited to easy preparation in the kitchen. Details of this method are available from the Australian Meat and Livestock Corporation.

Acknowledgements

The authors and publisher wish to acknowledge and thank the following for their assistance with reference material for this book:

Aus-Meat, Australian Meat and Live-Stock Corporation: pages 133–135

Australian Sheep and Wool Handbook

Department of Agriculture, NSW, Agfacts: pages 9–13, 72, 84

Elders Ltd

Harrington Agricultural Products Pty Ltd: page 96

Inkata Press; *Design of Shearing Sheds and Sheep Yards*, Barber and Freeman

Penguin Books Australia; *Continent in Crisis*, 1990: page 44

N. J. Phillips Pty Ltd: page 96

Rural Press Ltd: pages 54, 63, 97, 108, 109, 115, 116.

Index

Some other titles in the
PRACTICAL FARMING SERIES
Published by Inkata